Principles
of Chemical Sedimentology

Principles of Chemical Sedimentology

Robert A. Berner

*Associate Professor
of Geology and Geophysics
Yale University*

McGRAW-HILL BOOK COMPANY

New York	St. Louis	San Francisco
Düsseldorf	Johannesburg	Kuala Lumpur
London	Mexico	Montreal
New Delhi	Panama	Rio de Janeiro
Singapore	Sydney	Toronto

Principles of Chemical Sedimentology

Library of Congress Catalog Card Number 79-150775

07-004928-9

1234567890MAMM7987654321

This book was set in Times New Roman, and was printed
on permanent paper and bound by The Maple Press
Company. The designer was Marsha Cohen; the draw-
ings were done by BMA Associates, Inc. The editors
were Bradford Bayne and James W. Bradley. Stuart
Levine supervised production.

"Numberless in short are the ways, and sometimes imperceptible, in which the affections (emotions) colour and infect the understanding"

Francis Bacon
Novum Organum (1620)

Contents

Preface

The purpose of this book is to illustrate how sediments can be studied from a physicochemical viewpoint. By this I mean the application of chemical thermodynamics and kinetics to the elucidation of sedimentary problems. Sedimentary petrology for a long time has remained in a classical, descriptive stage; the intent here is to show how a completely different approach can be brought to bear on the study of sediments and sedimentary rocks. Because of this unorthodox approach, the reader will discover that this book is almost totally lacking in photomicrographs of thin sections, discussions of size distribution, cross section of sedimentary basins, classifications of rocks, etc., which are the characteristic features of most works on sedimentary petrology. In their place you will find such things as equilibrium calculations, diffusion equations, and diagenetic models. Discussion is generally confined to the more abundant inorganic sedimentary constituents which are involved in low-temperature chemical processes (e.g., chert, limestone, etc.). I hope that this book not only will show sedimentary petrologists the advantages of applying physical chemistry to their problems but also will show chemists and geochemists who may be interested in sediments the types of sedimentary problems which are amenable to experimental and theoretical treatment.

The subject matter presented here is based on two graduate courses in chemical sedimentology taught by the author over the past five years at Yale University. I am indebted to discussions with the various students who have forced me to clarify many cloudy points. Thoughtful comments and criticism of the manuscript, either in whole or in part, have been provided by the following individuals, whose help is gratefully acknowledged:

R. Siever of Harvard University, R. M. Garrels and J. Gieskes of the Scripps Institution of Oceanography, M. Lafon of the State University of New York at Binghamton, D. J. J. Kinsman of Princeton University, B. B. Hanshaw of the U.S. Geological Survey, I. R. Kaplan of the University of California at Los Angeles, D. L. Graf of the University of Illinois, and B. J. Skinner and E. A. Perry of Yale University. Of course the author is responsible for any ideas expressed, and this, in cases of disagreement, lets any or all of the above-mentioned off the hook. Finally, I thank my wife, Elizabeth K. Berner, for her help in preparing the index and in trying to make the text more readable, and for her patience while I have focused my attention for the past year almost entirely on writing and rewriting this book. Musical inspiration was provided by Maurice Ravel.

Robert A. Berner

List of Symbols

Below, listed in alphabetical order and followed by Greek and nonalphabetical type, are those symbols most frequently used in this book. All other symbols are found in close proximity to their definitions in the text. For certain fundamental constants, standard numerical values are cited.

Symbol	Definition
A	Surface area; surface area of a crystal
$\bar{A}$	Surface area of crystals per unit volume of water
A_c	Carbonate alkalinity
a	Activity
$\bar{a}$	Space-averaged activity within a Gouy layer
aq	Aqueous; refers to substances in solution
C, c	Concentration in aqueous solution in mass per unit volume
C_0, c_0	Concentration in a sediment pore water at $x = 0$
C_s, C_s	Concentration at the surface of a solid assumed to represent the equilibrium or saturation concentration
C_∞	Concentration in the bulk solution surrounding a dissolving or precipitating body at a distance large compared to the size of the body; limiting concentration in a pore water as $x \to \infty$
D	Diffusion coefficient (Fick's law)
D_s, D_s	Whole sediment diffusion coefficient
E	Reversible electrical potential (emf) of a redox reaction
E°	Standard-state electrical potential of a redox reaction
$\bar{E}$	Space-averaged electrical potential in a Gouy layer
Eh	Electrical potential of an oxidation half reaction relative to the standard hydrogen half reaction
F	Gibbs free energy (or simply free energy)
ΔF	Change in Gibbs free energy for a chemical reaction
ΔF°	ΔF when each product and reactant is in its standard state
$\Delta F \Gamma_f^\circ$	Standard free energy of formation from the elements
ΔF^*	Free energy of formation of the critical nucleus
$\mathscr{F}$	Faraday constant (23.06 kcal volt^{-1} equiv^{-1}; 96,494 coulombs equiv^{-1})
h	Partial molal enthalpy

Symbol	Definition
h°	Partial molal enthalpy in the standard state
Δh°_f	Standard enthalpy of formation from the elements
I	Ionic strength
IAP	Ion-activity product
IAP$^\circ$	Ion-activity product in supersaturated solution prior to precipitation
IMP	Ion molality product
J	Diffusion flux vector (mass/area/time)
J_x	Diffusion flux vector for a sediment (mass/area of sediment/time)
K	Thermodynamic, or activity equilibrium constant
K_m	Molality equilibrium constant
k	Generalized rate constant
k_l	First order rate constant
$\mathcal{K}$	Boltzmann constant (1.38×10^{-16} erg deg^{-1}; 3.30×10^{-27} kcal deg^{-1})
ln	Natural or base e logarithm (ln = 2.303 log)
log	Common or base 10 logarithm (log = 0.434 ln)
m	Molality (moles per kg of water)
m_T	Total molality; for a substance A this includes all dissolved free ions, ion pairs, and other forms of A; m_T can be written as m_{T_A} or m_{A_T}
$\bar{m}$	Space-averaged molality in a Gouy layer
$\bar{M}$	Mass per unit area
n	Number of equivalents of electrons transferred during a redox reaction; or number of moles
n_a	Number of atoms or ions in a crystal
P	Total pressure
P_g	Partial pressure (P with a subscript)
pH	$- \log a_{H^+}$
pK	$- \log K$
ppm	Parts per million (in dilute solution; mg/liter)
pS^{--}	$- \log a_{S^{--}}$
Q	Ratio of the activities of products to reactants for any chemical reaction

Symbol	Definition
$Q_{\gamma T}$	Product of total activity coefficient, (IAP/IMP) or (K/K_m)
R	Gas constant $(1.99 \times 10^{-3}$ kcal mole^{-1} deg^{-1}); or rate of deposition
r	Nominal radius of a crystal $(3V/A)$
r^*	Nominal radius of the critical nucleus
s	Partial molal entropy
s°	Standard-state entropy
T	Absolute temperature in degrees Kelvin $(= {}^\circ C + 273.15)$
t	Time
u	Flow velocity of ground water
v	Partial molal volume; volume per mole for pure phases, crystals, etc.
$v^\frown$	Partial molal volume in the standard state
V	Volume; volume of a crystal
X	Mole fraction
x	Depth in a sediment; generalized distance coordinate
Z	Valence charge of an ion
$\alpha_{i,j}$	Harned's rule coefficient
γ	Practical activity coefficient (a/m)
γ_T	Total activity coefficient (a/m_T)
$\gamma_\pm$	Mean activity coefficient of a salt
$\gamma_{\pm 1(0)}$	Mean activity coefficient for dissolved salt 1 in the presence of no other dissolved salts
$\gamma_{(0)\pm 1}$	Mean activity coefficient of dissolved salt 1 at zero concentration in the presence of other dissolved salts
$\bar{\gamma}$	Space averaged activity coefficient in a Gouy layer
θ	Sediment tortuosity
λ	Rational activity coefficient (a/X)
μ	Chemical potential
μ°	Chemical potential in the standard state
μ_{el}	Electrochemical potential
ρ	Radial coordinate
σ	Specific surface (interfacial) free energy
ΣCO_2	Total dissolved carbonate

Symbol	Definition
Φ	Porosity
Φ_0	Porosity at $x = 0$
ω	"Rate of deposition" (dx/dt)
$đ$	Density
$đ_s$	Average density of solids in a sediment
$đ_w$	Density of aqueous solution ("water")
$\oplus$	Degree of supersaturation $[RT\ln(\mathrm{IAP}_0/K)]$ or $[\mathcal{K}T\ln(\mathrm{IAP}_0/K)]$

1
Introduction

What is chemical sedimentology? Literally, chemical sedimentology is either the study of chemical sediments or the chemical study of sediments. The two definitions are quite different. The study of chemical sediments encompasses all aspects (e.g., age, composition, geologic setting, mode of formation) of sediments and sedimentary rocks which are classified in the traditional sense as being chemical. The alternative definition means a chemical approach to the study of sediments in general, regardless of how they are classified. The present book employs the latter definition. Discussion is directed toward those chemical processes which bring about the formation and alteration of some of the more common sedimentary minerals of low-temperature origin. The primary goal is to illustrate how the basic principles of physical chemistry can be applied to the solution of sedimentological problems; therefore, a physicochemical treatment of each subject is used wherever possible. This book is intended to serve as a theoretical counterpart to more descriptive and inclusive books devoted to sedimentary rocks. That is why the word "principles" appears in the title.

The study of sediments from a chemical viewpoint is often referred to as sedimentary geochemistry. This phrase is not used as a title for the present book because the approach is not that of a geochemist, but rather that of a sedimentary petrologist attempting to solve classical sedimentary problems.

PHYSICAL CHEMISTRY AND SEDIMENTS

Igneous rocks are uplifted into the zone of weathering. Molluscs secrete aragonite shells. Sea water, isolated from the ocean, contacts a dry air mass. Carbohydrates are produced by green plants during photosynthesis. These are some examples of how physical and biological processes bring about a state of chemical disequilibrium at the surface of the earth. In an attempt to reach equilibrium, chemical reactions take place. Thus, igneous minerals are weathered to clay minerals and dissolved silica, cations, and bicarbonate. Aragonite is converted to calcite during diagenesis. Evaporation of sea water brings about the precipitation of gypsum and halite. Carbohydrates are decomposed by bacteria in anaerobic sediments to methane and carbon dioxide. It is reactions such as these which make up the basic subject matter of this book. The purpose of the present section is to demonstrate the importance of physical chemistry to the problem.

One powerful tool for studying sedimentary chemical reactions is chemical thermodynamics. Thermodynamics enables prediction of the direction and extent of a reaction during the attainment of equilibrium. Consider the following example: An aragonite shell is buried in a sediment. Given enough time, will it dissolve? The question can be answered from a knowledge of the pH and concentrations of dissolved calcium and bicarbonate in the pore water. These enable thermodynamic calculations of whether or not the pore water is undersaturated with respect to aragonite. If undersaturated, the calculated degree of undersaturation, plus a knowledge of the relative amounts of pore water and aragonite, can then be used to determine whether the shell will completely dissolve. If it cannot dissolve completely, the water will equilibrate with aragonite and the equilibrium concentration of dissolved calcium can also be calculated. Comparison of the equilibrium concentration with that predicted by thermodynamics for calcite equilibrium under the same conditions would show that the value for calcite is lower. Thus, under these conditions aragonite eventually should dissolve and reprecipitate as calcite.

The application of chemical thermodynamics to sedimentary problems is a relatively recent phenomenon, although equilibrium studies of salt deposits were made more than 60 years ago by van't Hoff and coworkers (van't Hoff et al., 1903). An outstanding recent example is the book by Garrels and Christ (1965). Although considerably more work needs to be done in this field (e.g., calculation of ion activities in supersaline brines and the stability of

clay minerals in sea water), much has already been accomplished.   Thermodynamic data for many of the common minerals, ions, etc., in sediments and
sea water are available and summarized in the present book in Appendix I.

Another powerful tool for studying sedimentary reactions is chemical
kinetics.   Chemical kinetics enables prediction of the rates and mechanisms
by which equilibrium is approached.   In contrast to thermodynamics,
kinetics has been largely neglected by sedimentologists.   This is unfortunate,
because kinetics is of special importance at the geologically low temperatures
encountered in sedimentary environments.   Reaction rates are often slow and
as a result metastability is common.   For instance, many metastable authigenic
minerals form and persist in sediments because the more stable phases, for a
variety of reasons, crystallize very slowly.   Opaline silica forms instead of
quartz, aragonite instead of calcite, gypsum instead of anhydrite, and
limonitic goethite instead of hematite.   During diagenesis equilibrium is
approached, but the rates are variable and dependent upon such kinetic
phenomena as rates of dissolution, transport of dissolved constituents by
diffusion, nucleation, and crystal growth.   An added complication is that the
rates of some diagenetic reactions are dependent upon bacterial activity.
Examples are the formation of dissolved NH_4^+ and pyrite.   The study of
bacterial metabolism belongs properly to the field of microbial physiology,
but the overall rate of change of dissolved species outside the cell can in some
cases be treated in terms of simple chemical kinetics (see Chapter 7).

Theoretical calculations of sedimentary reaction rates are far more
difficult than thermodynamic equilibrium calculations.   For one thing there
is a relative lack of data, but, more importantly, adequate theoretical models
are often not available.   In this book attempts are made to formulate kinetic
models and to test them wherever there are available rate data for natural
waters and sediments.   In this respect, the study of early diagenesis is most
useful.   Depth in recent sediments is proportional to time (assuming continuous sedimentation) and the depth time scale can be determined by radiometric dating.   Thus, the distribution with depth of a sedimentary parameter
(e.g., dissolved sulfate) serves as a natural long-term experiment in sedimentary
chemical kinetics.

Many sediments are not easily described by quantitative kinetic models.
Thus, a considerable portion of this book consists of qualitative discussions
based on natural observations and laboratory experiments.   Chemical changes
which take place at considerable depth, or which occur over many millions of
years, can sometimes be followed by drilling into thick sequences of more or
less continuously deposited sediments.   An example is the reaction of
montmorillonite with illite to form mixed-layer clay, which takes place over
intervals of thousands of meters in the Tertiary sediments of the Gulf Coast
(see Chapter 9).   In the absence of drilling, long-term or deep diagenetic
reactions can be studied only by comparison of uplifted ancient sedimentary

rocks with modern sediments believed to represent equivalents of each rock prior to diagenesis.   If the ancient rocks do not conform to thermodynamic equilibrium predictions, answers must be sought through laboratory study of possible kinetic inhibition mechanisms.   In this way qualitative explanations can be obtained.   An example is the extremely slow transformation of aragonite to calcite during diagenesis in sea water.   Laboratory studies have concluded that at least one factor in this transformation is the presence of high concentrations of dissolved Mg^{++} in sea water, which inhibits the crystallization of low-magnesium calcite (see Chapter 8).

INDEPENDENT VERSUS DEPENDENT VARIABLES

The chemical reactions which are of greatest importance to sedimentologists are heterogeneous; i.e., they involve the transfer of matter from one phase to another.   The two major types are solid-water reactions and air-water reactions.   Of the two, only solid-water reactions are discussed in this book, and the solid is usually a mineral.   Because of heterogeneous reactions, the compositions of minerals and associated natural waters are interdependent. As an example consider a ground water from an area of chemical weathering. It has a dissolved silica concentration which reflects both the nature of the original silicate rocks and the type of weathering product, or clay mineral, formed.   Silica is leached from the original rock and its concentration in solution is dependent upon the rate of leaching.   In this way dissolved silica behaves as a dependent variable.   However, the nature of the clay mineral formed depends upon the silica concentration in the water (see Chapter 9) so that in this respect dissolved silica is an independent variable.   The choice of whether a given parameter is treated as an independent or a dependent variable depends upon which aspects of a given problem are emphasized.

Most of this book is concerned with sedimentary mineral formation. Thus, water composition is usually treated as an independent variable, and because of its quantitative significance, sea water receives major attention. In order to be able to predict what minerals can form from sea water (or dissolve in it), its thermodynamic properties need to be known.   Besides temperature and pressure, this includes the activities of the principal dissolved species. A detailed discussion of ion activities in sea water (and other natural solutions) is presented in Chapters 3 and 4 so that this may provide a background for succeeding discussions of sedimentary mineral formation.

It is sometimes useful to also treat sea water buried in sediments as a dependent variable.   This is because in sediments there is a large mass ratio of minerals to dissolved solids.   Small changes in the composition of some minerals can be seen as relatively large changes in the concentration of ions in solution.   For instance, dissolution of 0.1 percent $CaCO_3$ in a typical mud containing 50 percent interstitial sea water results in a 100 percent increase in dissolved Ca^{++}. Thus, interstitial water is a sensitive indicator of diagenesis, and small changes in pore water composition can be used to infer changes in mineral composition which are otherwise undetectable by the usual mineralogical methods. This is also true of bacterial reactions involving sedimentary organic matter. Low concentrations of bacterial metabolites, such as dissolved ammonia and H_2S, can be readily measured in pore waters, although they may represent only a small fraction of the total reduced nitrogen and sulfur in the sediment.

REFERENCES

Plus general works on sedimentary rocks and low-temperature geochemistry

Borchert, H., and Muir, R. O., 1964, *Salt deposits: The origin, metamorphism, and deformation of evaporites*, Van Nostrand, London, 338 p.

Carozzi, A. V., 1960, *Microscopic sedimentary petrography*, Wiley, New York, 485 p.

Degens, E. T., 1965, *Geochemistry of sediments*, Prentice-Hall, Englewood Cliffs, N.J., 342 p.

Folk, R. L., 1968, *Petrology of the sedimentary rocks*, Hemphill's Book Store, Austin, Texas, 170 p.

Garrels, R. M., and Christ, C. L., 1965, *Solutions, minerals, and equilibria*, Harper, New York, 450 p.

Hatch, F. H., and Rastall, R. H., 1965, *Petrology of the sedimentary rocks*, 4th rev. ed. (rev. by J. T. Greensmith), Thomas Murby, London, 408 p.

Krauskopf, K. B., 1967, *Introduction to geochemistry*, McGraw-Hill, New York, 706 p.

Milner, H. B., 1962, *Sedimentary petrography*, 4th rev. ed., Macmillan, New York, 715 p.

Pettijohn, F. J., 1957, *Sedimentary rocks*, 2d ed., Harper, New York, 718 p.

Stumm, W., and Morgan, J. J., 1970, *Aquatic chemistry*, Wiley, New York, 583 p.

Twenhofel, W. H., 1961, *Treatise on sedimentation*, 2d ed. (1932), Dover, New York, 926 p.

Van't Hoff, J. H., Armstrong, E. F., Hinrichsen, W., Weigert, F., and Just, G., 1903, Gips und Anhydrit, *Zeit. Phys. Chem.*, v. 45, pp. 257–306.

2
Review of Physical Chemistry

The purpose of this chapter is to provide a brief outline of some of the physico-chemical principles which are important to the study of chemical sediment-ology. Additional discussion of physical chemistry, necessary to a fuller understanding of diagenesis, is provided in Chapter 6. Emphasis in this chapter is on chemical equilibria and the thermodynamics of electrolyte solutions; crystal nucleation, growth, and dissolution; and colloid chemistry.

SOME USEFUL THERMODYNAMIC EQUATIONS

Equations are presented in this section which enable calculation of equilibrium constants for chemical reactions from thermodynamic data. Although the equations are stated as a matter of fact, they can be derived from more funda-mental thermodynamic expressions in a straightforward mathematical manner as shown in Appendix III. For a background discussion of the equations and other aspects of chemical thermodynamics the reader is referred to standard thermodynamic texts such as Lewis and Randall (1961) or Denbigh (1963) and to books which are devoted to the specific application of chemical thermodynamics to geology such as Garrels and Christ (1965), Helgeson (1964), and Broecker and Oversby (1970).

Chemical reactions can be written in the generalized form

$$bB + cC \rightarrow dD + gG$$

where D and G are referred to as products and B and C as reactants. The small letters denote the relative number of moles (gram molecular weight) of each reactant used up and of each product formed. A useful parameter which expresses the tendency for a chemical reaction to proceed is the change in Gibbs free energy ΔF. The value of ΔF is related to the thermodynamic concentration or activity of each of the reactants and products by the relation

$$\Delta F = \Delta F^\circ + RT \ln \frac{a_D{}^d a_G{}^g}{a_B{}^b a_C{}^c} \tag{1}$$

where ΔF = change in Gibbs free energy, F. (For an exact definition of F see Appendix III)

$\Delta F^\circ = \Delta F$ when each product and reactant is present at unit activity in some arbitrarily specified standard state

a = activity, a dimensionless parameter related to concentration (see below)

R = gas constant (1.99×10^{-3} kcal mole^{-1} deg^{-1})

T = absolute temperature, Kelvin

$\ln$ = natural logarithm

For the sake of brevity, let

$$Q = \frac{a_D{}^d a_G{}^g}{a_B{}^b a_C{}^c}$$

For a chemical reaction to proceed spontaneously as written, ΔF must be less than zero; i.e., reaction must result in a net decrease in free energy. If $\Delta F > 0$ as written, the reaction can only proceed backward from right to left. If $\Delta F = 0$, the reaction will not proceed in either direction, or in other words the products and reactants are at equilibrium. In this case the arrow symbol should be replaced by double arrows which denote equilibrium ($\leftrightharpoons$). If $\Delta F = 0$,

$$\Delta F^\circ = -RT \ln K \tag{2}$$

where $K = Q$ at equilibrium.

The parameter K is the thermodynamic equilibrium constant which relates activities of products to those of reactants at equilibrium for a given pressure and temperature.   Equation (2) is a fundamental relation upon which much of the thermodynamic calculation in this book is based.   At 25°C (298°K), the temperature for which free energy data are normally tabulated, Eq. (2) in terms of common (base 10) logarithms is

$$\Delta F° = -1.364 \log K \qquad (3)$$

where $\Delta F°$ is expressed in kilocalories per mole.

To calculate K for a chemical reaction, $\Delta F°$ must be known for the specified pressure and temperature.   If $\Delta F°$ is known only for another pressure and temperature (generally 1 atm and 25°C), K can be calculated via Eq. (2) and then corrected for changes in pressure and temperature using the equations

$$\left. \frac{d \log K}{dP} \right|_T = \frac{-\Delta v°}{2.3RT} \qquad (4)$$

$$\left. \frac{d \log K}{dT} \right|_P = \frac{\Delta h°}{2.3RT^2} \qquad (5)$$

where $v°$ = standard partial molal volume

$h°$ = standard partial molal enthalpy

$\Delta h°$ = change in $h°$ for the reaction analogous to $\Delta F°$

$\Delta v°$ = change in $v°$ for the reaction excluding values for any gaseous reactants or products

P = total pressure

$\log$ = base 10 logarithm

The partial molal quantities refer to instantaneous changes in volume or enthalpy of a solution phase per mole of added component.   For pure phases, whose composition does not vary, v and h are simply the volume and enthalpy per mole.   Enthalpy is the isobaric heat content; i.e., the change in enthalpy during a chemical reaction is equal to the heat given off or absorbed at constant pressure.

Like $\Delta F°$, both $\Delta h°$ and $\Delta v°$ are functions of temperature and pressure. However, for the small changes in P and T encountered in most sedimentary environments, they can be regarded as constants, thereby enabling direct integration of Eqs. (4) and (5).   Only for precise calculations where data are available (see Chapter 4), does the variability of $\Delta h°$ with temperature and $\Delta v°$ with pressure need be considered.

ACTIVITY, STANDARD STATES, AND THERMODYNAMIC DATA

The parameters bearing the superscript ° in the previous discussion refer to values in a standard state where the activity is equal to one.   In this section standard states are defined for the various classes of chemical compounds found in sediments, and activity is related to concentration by means of activity coefficients. The activity of a substance is an idealized or effective concentration that obeys thermodynamic equations such as Eq. (2). In general, as a substance is diluted in solution its activity decreases. In this manner activity is proportional to concentration; however, it is not directly proportional, because the proportionality factor (the activity coefficient) is itself a function of concentration. Much confusion can be avoided throughout the following discussion if it is kept in mind that the activity is dimensionless and that activity coefficients are thereby expressed in units of reciprocal concentration or reciprocal pressure.

SOLID SOLUTIONS

The standard state is the pure end member solid at the P and T of the reaction. The concentration unit used is mole fraction and the relation between activity and concentration is

$$a = \lambda X \tag{6}$$

where X = mole fraction of the pure end member (i.e., number of moles of a component divided by the total number of moles of all components in the solid solution)

λ = rational activity coefficient

An important property of both solid and liquid solutions is that the following limiting law is obeyed:

$$\lim_{X \to 1} \lambda = 1 \tag{7}$$

In other words, as the solid solution approaches the pure end member in composition, the activity becomes approximately equal to the mole fraction. For the pure-end-member composition ($X = 1$), activity and mole fraction are strictly equivalent and therefore the activity is equal to one. In this way the pure end member is the standard state. For the minerals discussed in this book it is generally safe to assume that if $X > 0.95$, $\lambda \approx 1$ (Garrels and Christ, 1965). For other compositions appropriate expressions for calculating λ are given later in this chapter under "Activities of Nonelectrolytes," page 21.

WATER IN AQUEOUS SOLUTION

The standard state is pure water at the P and T of the reaction. The concentration units and limiting law are the same as those used for solid solutions. An additional useful relation for water is that, at low total pressure,

$$a \approx \frac{P_{H_2O}}{P^\circ_{H_2O}} \tag{8}$$

where P_{H_2O} = vapor pressure of water in solution

$P^\circ_{H_2O}$ = vapor pressure of pure water at the same P and T (standard state)

From Eqs. (6) and (8)

$$P_{H_2O} = P^\circ_{H_2O} \lambda X \tag{9}$$

In the region near $X = 1$, where $\lambda \approx 1$, Eq. (9) reduces to

$$P_{H_2O} = P^\circ_{H_2O} X \tag{10}$$

This expression is commonly referred to as *Raoult's law*.

SOLUTES IN AQUEOUS SOLUTION

The standard state is an ideal unimolal solution, where activity equals molality, at the P and T of the reaction. The concentration unit is molality (or moles per 1000 grams of water), and

$$a = \gamma m \tag{11}$$

where γ = practical activity coefficient

m = molality

The limiting law is

$$\lim_{m \to 0} \gamma = 1 \tag{12}$$

Thus, as the solution becomes sufficiently dilute the activity becomes essentially equal to molality. The standard state is a hypothetical solution of one-molal concentration where the limiting law is obeyed. It is *not* the actual activity at a concentration of one molal and should not be confused with it.

Although at first sight it might appear that the choice of standard state for a dissolved solute is overly complex, it turns out that the convention adopted is more useful than that which employs the pure-end-member concept. The dissolved salts in natural waters are never present in mole fractions exceeding 0.2 so that compositions near that of the pure-end-member solute cannot be physically realized. By contrast most natural waters including sea water are sufficiently dilute that limiting law behavior, i.e., $a \approx m$, for undissociated or neutral dissolved species is obeyed. For electrolytes a greater degree of dilution is necessary for γ to approach one; however, univalent ions in most river and lake water have activities within 10 percent of m.

GASEOUS SOLUTIONS

The standard state is the pure ideal gas at one atmosphere total pressure and the temperature of the reaction. The concentration unit is partial pressure so that

$$a = \chi P_g$$
$$= \chi P X \qquad (13)$$

where P_g = partial pressure in atmospheres

χ = gas activity coefficient

X = mole fraction

P = total pressure in atmospheres

The limiting law is

$$\lim_{P \to 0} \chi = 1 \qquad (14)$$

At the relatively low total gas pressures discussed in this book the value of χ is very close to one, and thus a good approximation is

$$a = P_g \qquad (15)$$

Under the conditions where Eq. (15) is obeyed, the gas is described as a perfect gas whose partial pressure obeys the relation

$$P_g = \frac{RT}{v} \qquad (16)$$

where v = molar volume of the gas.

In this book the term partial pressure is also applied to volatile species in aqueous solution, which requires some explanation. The partial pressure of a dissolved volatile species is the value calculated for thermodynamic equilibrium with a hypothetical gas phase also containing the species. The gas phase may or may not be present and may or may not exhibit exchange equilibrium with the aqueous solution. For example, a natural water may contain dissolved oxygen at a known value of a_{O_2}. The value of P_{O_2} for the dissolved oxygen is related to a_{O_2} via the reaction

$$O_{2 \text{ gas}} \leftrightharpoons O_{2 \text{ aq}}$$

$$K = \frac{a_{O_2}}{P_{O_2}} \qquad (17)$$

If a_{O_2} is known, P_{O_2} can be calculated via Eq. (17) (and compared with measured P_{O_2} if a gas phase is actually present).

Table 2-1   Standard states, limiting laws, and activity-concentration relations for the four major classes of sedimentary substances

Class	Standard state	Activity-concentration relation	Limiting law
Solid solution	Pure solid at P and T of reaction	$a = \lambda X$	$\lim_{X \to 1} \lambda = 1$
Water in aqueous solution	Pure water at P and T of reaction	$a = \lambda X$	$\lim_{X \to 1} \lambda = 1$
Solute in aqueous solution	Ideal one molal solution at P and T of reaction	$a = \gamma m$	$\lim_{m \to 0} \gamma = 1$
Gaseous solution	Pure ideal gas ($X = 1$) at $P = 1$ atm and T of the reaction	$a = \chi P X$	$\lim_{P \to 0} \chi = 1$

A summary of standard states, limiting laws, and activity-concentration relations for the four major classes of sedimentary substances is shown in Table 2-1.

STANDARD–STATE THERMODYNAMIC DATA

In order to calculate values for the equilibrium constant K via Eqs. (2), (4), and (5), one needs to know values of $F°$, $h°$, and $v°$. Some values of $v°$ at 298°K and 1-atm total pressure are listed in Appendix I. Since only changes in $F°$ or $h°$ can be determined, a relative scale has been adopted for these parameters. By convention, the values of $F°$ and $h°$ for the pure elements in their most stable form are arbitrarily chosen as zero. Values of $F°$ and $h°$ for all other substances can then be tabulated as $\Delta F_f°$ and $\Delta h_f°$, the standard free energy and standard enthalpy of formation from the elements. Values of $\Delta F_f°$ and $\Delta h_f°$ at 298°K and 1-atm total pressure for common sedimentary substances are listed in Appendix I. Also included in Appendix I are values of standard entropy $s°$ based on the third law of thermodynamics. [For a discussion of entropy and the third law see Lewis and Randall (1961).]

For dissolved (aqueous) ions $\Delta F_f°$, $\Delta h_f°$, and $s°$ for H_{aq}^+ are arbitrarily set equal to zero so that $\Delta F_f°$, $\Delta h_f°$, and $s°$ for Cl_{aq}^- are equal to $\Delta F_f°$, $\Delta h_f°$, and $s°$ for HCl_{aq}. Values of the same parameters for Na_{aq}^+ are equal to those for $NaCl_{aq}$ minus those for Cl_{aq}^-. By similar reasoning values for all remaining ions are obtained. Values of $v°$ for ions are calculated by a similar process.

As an example of the use of Appendix I consider the reaction

$$2H_{aq}^+ + FeS_{mackinawite} \leftrightharpoons Fe_{aq}^{++} + H_2S_{gas}$$

The standard free energy change at 25°C and 1-atm total pressure is given as

$$\Delta F° = \Delta F_{f_{Fe^{++}}}° + \Delta F_{f_{H_2S}}° - (2\Delta F_{f_{H^+}}° + \Delta F_{f_{FeS}}°)$$

Substituting values from Appendix I,

$$\Delta F^\circ = -5.89 \text{ kcal/mole}$$

From Eq. (3)

$$K = 10^{-\Delta F^\circ/1.364} = 10^{4.31} = \frac{a_{Fe^{++}} P_{H_2S}}{a_{H^+}^2 a_{FeS}}$$

Since mackinawite shows little solid substitution, $a_{FeS} \approx 1$. Thus

$$\frac{a_{Fe^{++}} P_{H_2S}}{a_{H^+}^2} = 10^{4.31}$$

REDOX REACTIONS AND Eh

An oxidation-reduction, or redox, reaction is one that involves the transfer of electrons between products and reactants. It can be written in the general form

$$bB_{red} + cC_{oxid} \rightarrow dD_{oxid} + gG_{red}$$

To express the number of equivalents of electrons n transferred, this reaction can be split up into two half reactions:

$$bB_{red} \rightarrow dD_{oxid} + ne \qquad \text{(half reaction I)}$$

$$gG_{red} \rightarrow cC_{oxid} + ne \qquad \text{(half reaction II)}$$

where the symbol e represents an electron.

The difference (reaction I − reaction II) equals the whole redox reaction. For each half reaction a standard free-energy change can be calculated using the arbitrary convention that ΔF_f° for an electron is equal to zero, i.e., for half reaction I, $\Delta F_I^\circ = d\Delta F_{f_D}^\circ - b\Delta F_{f_B}^\circ$ and for reaction II, $\Delta F_{II}^\circ = c\Delta F_{f_C}^\circ - g\Delta F_{f_G}^\circ$. The standard free-energy change for the whole redox reaction then is

$$\Delta F^\circ = \Delta F_I^\circ - \Delta F_{II}^\circ \tag{18}$$

Now if $C = H_{aq}^+$ and $G = H_{2\,gas}$, half reaction II is

$$gH_{2\,gas} \rightarrow 2gH_{aq}^+ + 2ge$$

where g is adjusted so that $2g = n$ for half reaction I. Since the standard free energies of formation of both H_{aq}^+ and $H_{2\,gas}$ are equal to zero, the free-energy change $\Delta F_{H_2-H^+}^\circ = 0$, and from Eq. (18)

$$\Delta F^\circ = \Delta F_I^\circ - \Delta F_{H_2-H^+}^\circ = \Delta F_I^\circ \tag{19}$$

Thus, from Eq. (1), for a redox reaction involving the hydrogen half reaction,

$$\Delta F = \Delta F_I^\circ + RT \ln \frac{a_D^d P_{H_2}^g}{a_B^b a_{H^+}^{2g}} \tag{20}$$

The free-energy change for a redox reaction is related to the reversible electrical potential developed if the reaction were to proceed in an electrochemical cell by the relations

$$\Delta F = n\mathscr{F}E \tag{21}$$

$$\Delta F^\circ = n\mathscr{F}E^\circ \tag{22}$$

where E = reversible potential (emf) in volts of the electrochemical cell

E° = potential if all products and reactants are in their standard states

n = number of equivalents of electrons transferred during the reaction

$\mathscr{F}$ = the Faraday constant (23.06 kcal volt^{-1} equiv^{-1})

From Eqs. (20), (21), and (22) for redox reactions involving the hydrogen half reaction,

$$E = E_1^\circ + \frac{RT}{n\mathscr{F}} \ln \frac{a_D{}^d P_{H_2}{}^g}{a_B{}^b a_{H^+}{}^{2g}} \tag{23}$$

We are now in a position to define the parameter Eh. The Eh is simply E for Eq. (23) when $P_{H_2} = 1$ and $a_{H^+} = 1$. In other words, the Eh of a half reaction is the calculated emf for an electrochemical cell where the half reaction is combined with the standard state $H_2 - H^+$ half reaction. Therefore

$$Eh_1 = E_1^\circ + \frac{RT}{n\mathscr{F}} \ln \frac{a_D{}^d \text{(oxid)}}{a_B{}^b \text{(red)}} \tag{24}$$

In Eq. (24), E° is calculated from ΔF° for the half reaction written as an oxidation, i.e., with the electrons to the right; as a result, Eh is sometimes referred to as the oxidation potential. Note that the activity of electrons does not appear in the activity quotient. Also, the activities of $H_{aq}{}^+$ and/or $H_{2\,gas}$ may appear in Eq. (24) as variables. This is because hydrogen ion or hydrogen gas at activities other than one can generate an emf when combined with the *standard state* $H_2 - H^+$ half reaction. For example, consider the half reaction

$$2NH_{3\,aq} \rightarrow N_{2\,gas} + 6H_{aq}{}^+ + 6e$$

whose Eh is given by

$$Eh = E^\circ + \frac{RT}{6\mathscr{F}} \ln \frac{a_{N_2} a_{H^+}{}^6}{a_{NH_3}^2} \tag{25}$$

Here, a_{H^+} is a genuine variable and not to be confused with $a_{H^+} = 1$ in the implicit $H_2 - H^+$ half reaction.

From the definition of Eh it can be seen that a half reaction which has a positive Eh will oxidize H_2 gas at a standard activity of one to H^+ ions also at an activity of one, whereas a half reaction with a negative Eh will reduce H^+ ions to hydrogen gas when both are in their standard state. More importantly, a half reaction with a lower Eh will proceed as written (undergo oxidation) when combined with a half reaction of higher Eh. Thus, half reactions of high Eh are said to be "oxidizing" and those of low Eh to be "reducing."

The Eh of a solution can in many instances be measured directly by using an inert platinum electrode in combination with a standard hydrogen electrode or a reference electrode of known fixed Eh. In many natural waters, however, electrochemical problems preclude accurate Eh measurement. This is discussed in detail in Chapter 7. Because of measurement difficulties, another parameter $p(e)$ is often used to express the redox state of a solution. It is defined as the negative logarithm of the activity of an electron and its value is directly related to Eh. However, because problems in the interpretation of electrode measurements are discussed in this book, it was decided that Eh would be a more useful measure of redox state than $p(e)$, which is not a directly measurable quantity.

ION ACTIVITIES

Although on a macroscopic scale single-ion free energies of formation and activities cannot be determined, the concept of individual ion activity is still useful, and theoretical calculations and operational definitions of ion activity can be employed. The standard-state and activity-concentration relationship is the same as that for a neutral dissolved solute:

$$a_+ = \gamma_+ m_+ \tag{26}$$

$$a_- = \gamma_- m_- \tag{27}$$

where $_+$ refers to cations and $_-$ to anions. The individual ion-activity coefficient γ_+ or γ_- can be calculated from theory and the theory checked with measurements by combining γ_+ with γ_- to calculate the activity coefficient for a dissolved salt $\gamma_\pm$. For a dissolved salt $A_x B_y$ which dissociates into x cations and y anions

$$\gamma_\pm = (\gamma_+{}^x \gamma_-{}^y)^{1/(x+y)} \tag{28}$$

Values of $\gamma_\pm$ have been measured for a variety of salt solutions, and a good compendium of results is given in Robinson and Stokes (1959).

The most commonly used expression for calculating individual ion-activity coefficients is the Debye-Hückel equation. The assumptions behind the equation are: (1) deviations from ideal-solution behavior are due entirely to coulombic interactions between oppositely charged ions; (2) ions of opposite charge surround a central ion with spherical symmetry and the charge density is related to electrical potential by a linear model. This gives rise to the so-called ion atmosphere. The linear model is valid only if electrical interactions are over considerable distances. Therefore, the theory becomes more valid as the solution becomes more dilute.   The excess electrical free energy calculated from the model is equated to $RT \ln \gamma$ and the resulting equation is

$$\log \gamma = \frac{-AZ^2\sqrt{I}}{1 + \mathring{a}B\sqrt{I}} \tag{29}$$

where $\log$ = base 10 logarithm

$\quad\quad\quad Z$ = valence charge of the ion

$\quad\quad\quad I$ = ionic strength (see below)

$\quad\quad\quad A, B$ = constants, depending on temperature and nature of the solvent

$\quad\quad\quad \mathring{a}$ = "distance of closest approach" of ions and is crudely a measure of the radius of the hydrated ion

Values of A and B for water at various temperatures and $\mathring{a}$ are given in Appendix II.   Ionic strength is defined as

$$I = \tfrac{1}{2} \sum m_i Z_i^2 \tag{30}$$

where m_i refers to each and every ion in solution.

Tests of the Debye-Hückel equation (29) using equations of the type (28) have shown it to be reliable for predicted activities for free ions (those which do not undergo association or complex formation) over ionic strengths ranging up to about 0.1.   At higher ionic strengths no strictly a priori theoretical models, especially for mixed electrolytes as encountered in natural waters, are available.   One simple method which has been shown to be reasonably accurate for sea water at $I \approx 0.7$ (see Chapter 3) is the *mean-salt method*.

The mean-salt method makes use of Eq. (28) and the following assumptions:

1. $\gamma_{K^+} = \gamma_{Cl^-}$ in pure KCl solution.
2. γ_{K^+} or γ_{Cl^-} in any given solution has the same value as γ_{K^+} or γ_{Cl^-} in a KCl solution of the same ionic strength.
3. γ values for other ions have the same value as γ values in pure potassium or chloride salt solutions of each ion.

As an example of this method consider $\gamma_{Mg^{++}}$ in a 0.1 m MgBr$_2$ solution. By the above method

$$\gamma_{Mg^{++}_{0.1\ m\ MgBr_2}} = \gamma_{Mg^{++}_{0.1\ m\ MgCl_2}}$$

$$\gamma_{Cl^-_{0.1\ m\ MgCl_2}} = \gamma_{Cl^-_{0.3\ m\ KCl}}$$

$$\gamma_{Cl^-_{0.3\ m\ KCl}} = \gamma_{K^+_{0.3\ m\ KCl}}$$

From Eq. (28)

$$\left(\gamma_{Mg^{++}}\gamma_{Cl^-}^2\right)^{\frac{1}{3}} = \gamma_{\pm MgCl_2}$$

$$\left(\gamma_{K^+}\gamma_{Cl^-}\right)^{\frac{1}{2}} = \gamma_{\pm KCl}$$

Thus

$$\gamma_{Mg^{++}_{0.1\,m\,MgBr_2}} = \frac{\gamma_{\pm 0.1\,m\,MgCl_2}^3}{\gamma_{\pm 0.3\,m\,KCl}^2}$$

This method is quite useful and reasonably accurate for moderately saline natural waters, but only if the ion under consideration does not undergo complex formation. After correcting for ion-pair or complex formation the method can still be used provided that the bridging solution, i.e., $MgCl_2$ above, does not itself exhibit ion-pairing. Values of ion-activity coefficients calculated by the mean salt method and Debye-Hückel equation are compared in Fig. 2-1.

For more concentrated solutions ($I > 1.0$) the mean-salt method can be improved by correcting for the effects of the other dissolved salts, in a mixed electrolyte, on $\gamma_\pm$ for each salt. This is most easily done via the empirical equation known as Harned's rule (Robinson and Stokes, 1955; Harned and Owen, 1958), which is

$$\log\gamma_{\pm 1} = \log\gamma_{\pm 1(0)} + \alpha_{1,2}I_2 \tag{31}$$

where $\gamma_{\pm 1}$ = mean activity coefficient of electrolyte 1 in the presence of another electrolyte 2

$\gamma_{\pm 1(0)}$ = mean activity coefficient in a solution of pure electrolyte 1 with an ionic strength the same as the total ionic strength ($I_1 + I_2$) of the electrolyte mixture

I_2 = ionic strength contributed by electrolyte 2 (assuming complete dissociation)

$\alpha_{1,2}$ = Harned's rule coefficient. Note that this refers to the effect of salt 2 upon salt 1 and not vice versa. In general, $\alpha_{1,2} \neq \alpha_{2,1}$

In the limiting case where $I_1 = 0$ and $I_2 = I_{total}$

$$\log\gamma_{(0)\pm 1} = \log\gamma_{\pm 1(0)} + \alpha_{1,2}I_{total} \tag{32}$$

where $\gamma_{(0)\pm 1} = \gamma_{\pm 1}$ when $I_1 = 0$.

As an example of the use of Harned's rule consider the previous example of $\gamma_{Mg^{++}}$ in $MgBr_2$ solution, but now assume that $m_{MgBr_2} = 1$ ($I = 3$). In this case

$$\log\gamma_{(0)\pm MgCl_2} = \log\gamma_{\pm 1.0\,m\,MgCl_2} + 3\alpha_{MgCl_2;\,MgBr_2}$$

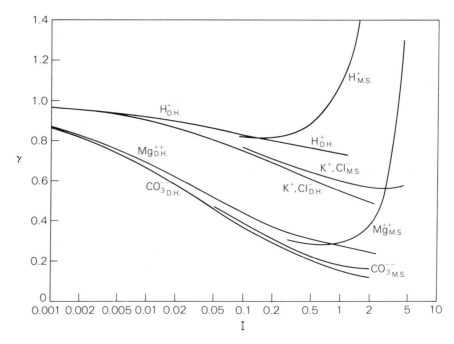

Fig. 2-1 Ion-activity coefficients (γ) as a function of ionic strength (I) calculated by the Debye-Hückel equation and by the mean-salt method, 25°C. M.S. = mean salt, D.H. = Debye-Hückel.

$$\gamma_{Mg^{++}_{M.S.}} = \gamma^{3}_{\pm MgCl_2}/\gamma^{2}_{\pm KCl} \qquad \gamma_{H^{+}_{M.S.}} = \gamma^{2}_{\pm HCl}/\gamma_{\pm KCl}$$

$$\gamma_{CO_3^{--}_{M.S.}} = \gamma^{3}_{\pm K_2CO_3}/\gamma^{2}_{\pm KCl} \qquad \gamma_{K^{+}_{M.S.}} = \gamma_{Cl^{-}_{M.S.}} = \gamma_{\pm KCl}$$

Note the disagreement between the mean salt and Debye-Hückel methods above $I = 0.2$. Divergence of the Debye-Hückel curves from one another with increasing ionic strength is due to differences in valence of the ions, and differences in the ion-size parameter $\mathring{a}$. (*Data from Robinson and Stokes, 1959, and Garrels and Christ, 1965.*)

and

$$\gamma_{Mg^{++}} = \frac{\gamma^{3}_{(0)\pm MgCl_2}}{\gamma^{2}_{\pm KCl}}$$

With this correction for $\gamma_{\pm MgCl_2}$ the third assumption remaining in the mean-salt method is unnecessary. Harned's rule has been shown to be valid for a large number of two-electrolyte mixtures at ionic strengths from 1 to greater than 10 (Harned and Owen, 1958).

For more than two dissolved salts, e.g., NaCl–MgCl₂–KCl, Harned's rule can be extended in the form

$$\log \gamma_{\pm_1} = \log \gamma_{\pm_{1(0)}} + \Sigma \alpha_{1,i} I_i \qquad (33)$$

where I_i = ionic strength for each salt other than salt 1

$\alpha_{1,i}$ = Harned's rule coefficient for the effect of each salt (in two-salt mixtures) on salt 1

Equation (33) assumes that the experimental values of $\alpha_{1,i}$ from studies of two-salt mixtures can be used for higher number salt mixtures; i.e., α is independent of composition. The work of Åkerlöf (1934) suggests that this assumption is reasonably valid. Values of $\alpha_{1,i}$ can be found in Chapter 3 in conjunction with the discussion of ion activities in supersaline brines.

The most accurate method for obtaining ion-activity coefficients, especially at high ionic strength, is through the use of ion-sensitive electrodes. The great advantage of electrodes is that they measure the activity of the ion directly and there is no need to correct for ion-pairing, etc. Again, however, the measured activity is an operational parameter defined in terms of a standard. The working equation for an ideal electrode is

$$\ln a = \frac{(E - E_{std})n\mathscr{F}}{RT} + \ln a_{std} \tag{34}$$

Thus a is defined in terms of a_{std}, the activity in a standard solution. Various standard solutions can be used. The pH scale is based on a primary acid solution which is dilute enough to enable calculation of ion activity by means of the Debye-Hückel equation. Secondary standards, known as buffer solutions, are calibrated against the primary standard and used in everyday pH determinations. For other cations it is often useful to use chloride solutions which exhibit no ion-pairing as primary and working standards. It is best to choose standard solutions as similar as possible to the natural waters in ionic strength and composition so as to minimize errors due to differences in liquid-junction potential at the surface of the reference electrode. (For further discussion of liquid-junction potentials see Ives and Janz, 1961.) For more concentrated waters this means that the standard solution may well fall outside the range of applicability of the Debye-Hückel equation. In this case either Eq. (28) or some arbitrary definition of a_{std} must be used to calculate the single-ion activity. If only a relative comparison of ion activities in different natural waters is sought, then any arbitrary a_{std} value can be used.

At the time of writing of this book specific ion electrodes were available for measuring activities of H^+, Na^+, K^+, Cl^-, F^-, S^{--}, and NO_3^- in natural waters with little or no unavoidable interference by other ions. Electrodes for measuring Ca^{++} and Mg^{++} were also available, but the electrodes showed some interference by other ions commonly encountered in natural waters.

COMPLEXING AND ION ASSOCIATION

Oppositely charged ions or ions and polar molecules will often interact with one another much more strongly than predicted by the Debye-Hückel equation.   The strength of the interaction is most simply represented by an equilibrium constant.   For example,

$$Mg_{aq}^{++} + SO_{4\,aq}^{--} \leftrightharpoons MgSO_{4\,aq}^{0}$$

$$K = 10^{2.3} = \frac{a_{MgSO_4^0}}{a_{Mg^{++}} a_{SO_4^{--}}}$$

$$Fe_{aq}^{3+} + H_2O_{liq} \leftrightharpoons FeOH_{aq}^{++} + H_{aq}^{+}$$

$$K = 10^{2.4} = \frac{a_{Fe(OH)^{++}} a_{H^+}}{a_{Fe^{3+}}}$$

Thus, the strong interaction is treated as an independent dissolved species with its own activity and activity coefficient.   The concentration of an un-complexed or free ion is calculated from its total or analytical concentration by subtracting from the total the sum of concentrations of all ion pairs or complexes:

$$m = m_T - \Sigma m_{complex} \tag{35}$$

where m = molality of free ion

$\quad\quad m_T$ = total (analytical) molality

Once the value of m is determined, then activity can be calculated using the methods of the previous section.

Some general rules are useful in considering ion-complexing in solutions:

1. The higher the charge of an ion, the more likely it is to form a strong (and abundant) complex.
2. The higher the concentration of a complexing species, the greater the degree of complexing.
3. No appreciable ion association is found between Cl^- and the alkali and alkaline earth cations at sedimentary temperatures.
4. Trivalent (and higher valent) ions are invariably associated with OH^- (or O^{--}) at neutral pH; i.e., they react with water.
5. Neutral organic molecules often form strong complexes with transition metal ions.

Occasionally it is useful to lump the effects of complexing with normal electrostatic attraction, etc., into an empirical or total activity coefficient γ_T which is defined as

$$\gamma_T = \frac{a}{m_T} \tag{36}$$

Since $a = \gamma m$

$$\gamma_T = \frac{m}{m_T} \gamma \qquad (37)$$

Values of γ_T can be determined directly by the measurement of a with electrodes and m_T by the usual analytical procedures. Knowledge of γ_T (and γ) enables the calculation of m and thus the degree of association $(m_T - m)/m_T$.

ACTIVITIES OF NONELECTROLYTES

Electrolyte solutions, in addition to dissolved ions, may also contain undissociated neutral species which include dissolved gases, neutral ion pairs, organic molecules, and undissociated weak acids. Experimentally as a good first approximation it has been found that the activity coefficient of neutral dissolved species can be represented by the empirical equation

$$\log \gamma = k_m I \qquad (38)$$

where k_m = salting coefficient

I = ionic strength

The value of k_m for most undissociated substances is greater than one. This leads to the term *salting out* when referring to the decrease in equilibrium molality or solubility of many neutral species with increasing salinity. Values of k_m in a large variety of electrolyte solutions are given in Harned and Owen (1958).

When dealing with ion-exchange and solid solutions the concepts of ideal and regular solution are useful. An ideal solution is one in which the activity coefficients of all components at all concentrations are equal to unity; i.e., activity is identical with mole fraction. In actuality, there are no strictly ideal solutions, although some real solutions approximate ideal behavior. Regular solutions, as defined in this book, are those whose rational activity coefficients λ follow the reciprocal relations

$$\ln \lambda_i = AX_j^2 + B(1 - 4X_i) X_j^2 \qquad (39)$$

$$\ln \lambda_j = AX_i^2 - B(1 - 4X_j) X_i^2 \qquad (40)$$

where i, j refer to the two components of a binary solution

$$X_i = 1 - X_j$$

The parameters A and B are related to thermodynamic variables but for present purposes can be considered simply as empirical coefficients. When $B = 0$, the regular solution is described as being symmetrical. Otherwise it is asymmetrical.

SOLUBILITY PRODUCTS AND SATURATION

For a reaction of the type

$$A_x B_{y\,\text{solid}} \rightleftharpoons xA_{\text{aq}}^{+y} + yB_{\text{aq}}^{+x}$$

the equilibrium constant K is

$$K = \frac{a_A{}^x a_B{}^y}{a_{A_x B_y}} \qquad (41)$$

If $A_x B_y$ is essentially a pure solid, its activity ≈ 1, and (41) reduces to

$$K = a_A{}^x a_B{}^y \qquad (42)$$

An equation of this sort is known as the *ion-activity solubility product* for the solid. It is the value measured at equilibrium or calculated by use of Eq. (2) from free-energy data. The actual ion-activity product of a given solution is denoted by IAP and may not be the same as K. The state of saturation of a solution with respect to a solid is defined as

$$
\begin{aligned}
&\text{IAP} > K \qquad \text{supersaturated} \\
&\text{IAP} = K \qquad \text{saturated} \qquad (43) \\
&\text{IAP} < K \qquad \text{undersaturated}
\end{aligned}
$$

The expression for K can be rewritten in terms of total-activity coefficients and total (analytical) concentrations:

$$K = \gamma_{T_A}{}^x \gamma_{T_B}{}^y m_{T_A}{}^x m_{T_B}{}^y = Q_{\gamma_T} K_m \qquad (44)$$

where $Q_{\gamma_T} = \gamma_{T_A}{}^x \gamma_{T_B}{}^y$

$$K_m = m_{T_A}{}^x m_{T_B}{}^y$$

K_m, the ion-molal solubility product, is a function not only of P and T but also of ionic strength and chemical composition. Corresponding to K_m is IMP, the actual ion-molal product. IMP and K_m follow the same relationship as K and IAP above. Also

$$\text{IAP} = Q_{\gamma_T} \text{IMP} \qquad (45)$$

Equations of the type

$$K = Q_{\gamma_T} K_m$$

apply not only to solubility products but also to all equilibrium constants. In order to calculate the effect of pressure and temperature upon K_m, the variability of γ_T with P and T must be known. The correct equations are

$$\left. \frac{d\log \gamma_T}{dP} \right|_T = \frac{v - v^\circ}{2.3RT} \qquad (46)$$

$$\frac{d\log \gamma_T}{dT}\bigg|_P = \frac{h^\circ - h}{2.3RT^2} \tag{47}$$

where v, h = partial molal volume and enthalpy in the solution under consideration

$v^\circ, h^\circ = v$ and h in the standard state (ideal one-molal solution)

From Eq. (44)

$$\frac{d\log K_m}{dP}\bigg|_T = \frac{d\log K}{dP}\bigg|_T - \frac{d\log Q_{\gamma_T}}{dP}\bigg|_T \tag{48}$$

$$\frac{d\log K_m}{dT}\bigg|_P = \frac{d\log K}{dT}\bigg|_P - \frac{d\log Q_{\gamma_T}}{dT}\bigg|_P \tag{49}$$

From Eqs. (46) through (49) and Eqs. (4) and (5), the effect of P and T upon K_m is

$$\frac{d\log K_m}{dP}\bigg|_T = \frac{-\Delta v}{2.3RT} \tag{50}$$

$$\frac{d\log K_m}{dT}\bigg|_P = \frac{\Delta h}{2.3RT^2} \tag{51}$$

HOMOGENEOUS KINETICS

The rate and mechanism of chemical reactions occurring within a single phase is the subject of homogeneous kinetics. Most homogeneous reactions in natural waters are so rapid that equilibrium is usually already present and study of rates from a geochemical standpoint is uninteresting. Outstanding exceptions occur, however, such as the inorganic reduction of sulfate and the thermal degradation of amino acids. Radioactive decay within sediments also can be treated in terms of homogeneous kinetics. Discussion here is brief and strictly phenomenological.

Chemical reactions can be classified according to the rate law which they follow. The simplest examples are

1. *Zeroth order:*

$$\frac{dC}{dt} = -k_0 \tag{52}$$

where C = concentration

t = time

Integrating,

$$C_0 - C = k_0 t \tag{53}$$

where $C_0 = C$ at $t = 0$

k_0 = zeroth-order rate constant

2. *First order:*

$$\frac{dC}{dt} = -k_1 C \tag{54}$$

Integrating,

$$C = C_0 \exp(-k_1 t)$$

where k_1 = first-order rate constant.

Radioactive decay and rate-controlling elementary chemical reactions of the type

$$A \rightarrow B + G \quad \text{or} \quad A \rightarrow B$$

follow this rate law.

3. *Second order:*

$$\frac{\partial C_A}{\partial t} = -k_2 C_A C_B \tag{55}$$

or

$$\frac{\partial C}{\partial t} = -k_2 C^2 \tag{56}$$

Equation (55) is solved by substituting $C_A = C_{A_0} - \Delta$ and $C_B = C_{B_0} - \Delta$, where Δ is a reaction variable. The solution is

$$\frac{1}{C_{A_0} - C_{B_0}} \ln \frac{C_A}{(C_{A_0} - C_{B_0}) - C_A} = k_2 t + \text{const} \tag{57}$$

Integration of Eq. (56) yields

$$\frac{1}{C} - \frac{1}{C_0} = k_2 t \tag{58}$$

The change in rate constant with temperature follows a relation similar to that for the equilibrium constant:

$$\frac{d \ln k_i}{dT} = \frac{U_a}{RT^2} \tag{59}$$

where U_a = activation energy.

Upon integration at constant U_a

$$k_i = A \exp(-U_a/RT) \tag{60}$$

where A = "frequency factor."

Equation (60) is known as the *Arrhenius equation*, and the term exp $(-U_a/RT)$ is identified as the Boltzmann expression for the fraction of species which have energies in excess of U_a.   Most reactions are described by this expression.

DIFFUSION

The diffusion of dissolved species in aqueous solution can be described by the equations

$$J = -D\,\text{grad}\,C \tag{61}$$

$$\frac{\partial C}{\partial t} = -\text{div}\,J = \text{div}\,(D\,\text{grad}\,C) \tag{62}$$

where J = diffusion flux vector (in mass per unit area per unit time)

C = concentration (in mass per unit volume)

D = diffusion coefficient

t = time

grad = gradient vector (∇)

div = divergence ($\nabla\cdot$)

At constant D, the one-dimensional (x-direction) forms of Eqs. (61) and (62) reduce to

$$J = -D\frac{\partial C}{\partial x} \tag{63}$$

$$\frac{\partial C}{\partial t} = D\frac{\partial^2 C}{\partial x^2} \tag{64}$$

Equations (63) and (64) are known as *Fick's first and second laws of diffusion*, respectively.

A geometric derivation of Fick's second law can be seen by reference to Fig. 2-2.   Consider two parallel planes perpendicular to the x axis located at x and $x + \Delta x$.   Diffusion across each plane is governed by Fick's first law as follows:

$$J_x = -D\frac{\partial C}{\partial x}\bigg|_x$$

$$J_{x+\Delta x} = -D\frac{\partial C}{\partial x}\bigg|_{x+\Delta x} \tag{65}$$

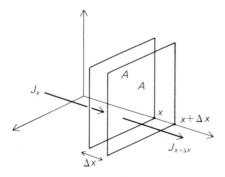

Fig. 2-2 One-dimensional diffusion model for the derivation of Fick's second law of diffusion.

The rate at which dissolved mass M changes in the volume element bounded by Δx and the area A is the difference between the rates of influx and outflux:

$$\frac{\Delta M}{\Delta t} = -A(J_{x+\Delta x} - J_x) \tag{66}$$

Similarly the rate of concentration change in the same volume element, since $\Delta M = A \, \Delta x \, \Delta C$, is

$$\frac{\Delta C}{\Delta t} = \frac{-(J_{x+\Delta x} - J_x)}{\Delta x} \tag{67}$$

From Eq. (65)

$$\frac{\Delta C}{\Delta t} = \frac{D[(\partial C/\partial x)_{x+\Delta x} - (\partial C/\partial x)_x]}{\Delta x} \tag{68}$$

Thus, passing to the limit $\Delta x \to 0$,

$$\frac{\partial C}{\partial t} = D\frac{\partial^2 C}{\partial x^2} \tag{64}$$

In Chapter 6 one-dimensional diffusion is discussed in further detail as it relates to sediment diagenesis. For spherically symmetric diffusion, which is useful when discussing crystal growth and dissolution, Eqs. (61) and (62) are

$$J = -D\frac{\partial C}{\partial \rho} \tag{69}$$

$$\frac{\partial C}{\partial t} = D\frac{\partial^2 C}{\partial \rho^2} + \frac{2}{\rho}\frac{\partial C}{\partial \rho} \tag{70}$$

where ρ = radial distance from the center of symmetry.

ENERGETICS OF NUCLEATION

Ions or atoms at the surface of a solid are not completely surrounded by other ions and atoms and as a result are not bound to the solid as strongly as they would be if they were present in the interior of the solid. As a result the surface ions and atoms have an excess surface free energy which is denoted by F_{surf} and defined as the energy required to move an ion or atom from the center of a crystal to its surface. For "large" crystals, larger than one micron, the bulk free energy F_{bulk} is much greater than F_{surf} so that F_{surf} can be neglected in thermodynamic calculations. During initial crystallization, however, the crystal embryos are very tiny and as a result have a high surface free energy which cannot be neglected. Thus, surface free energy is an important aspect of nucleation and crystal growth.

The precipitation of a pure solid from electrolyte solution can be represented by

$$x A_{aq}{}^+ + y B_{aq}{}^- \rightarrow A_x B_{y\,solid}$$

The total free-energy change for the reaction is

$$\Delta F = \Delta F_{bulk} + \Delta F_{surf} \tag{71}$$

From Eq. (1)

$$\Delta F_{bulk} = -nRT \ln \frac{IAP_0}{K} \tag{72}$$

where IAP_0 = ion-activity product in the original supersaturated solution

K = equilibrium ion-activity product

n = number of moles precipitated

Introducing the definitions

$$\sigma = \frac{dF_{surf}}{dA} \tag{73}$$

where σ = specific surface free energy for the interface between crystals and water

A = surface area

and

$$\oplus = RT \ln (IAP_0/K) \tag{74}$$

Eq. (71) at constant σ becomes (Nielsen, 1964)

$$\Delta F = -n\oplus + \sigma A \tag{75}$$

If Eq. (75) refers to the precipitation of one crystal, n is replaced by n_a, the number of atoms or ions rather than moles, and R is replaced by $\mathscr{K}$ (the Boltzmann constant) in the expression for $\oplus$. Also

$$V = n_a v_a \tag{76}$$

$$A = bV^{\frac{2}{3}} \tag{77}$$

where v_a = volume of an atom or ion in the crystal

  n_a = number of ions or atoms in a crystal

  b = geometric constant relating area to volume of the crystal

  V = volume of crystal

  A = area of crystal

From Eqs. (75), (76), and (77)

$$\Delta F = -n_a \oplus + \sigma b v_a^{\frac{2}{3}} n_a^{\frac{2}{3}} \tag{78}$$

A plot of Eq. (78) for arbitrary values of $\oplus$ and $\sigma b v_a^{\frac{2}{3}}$ is presented in Fig. 2-3. Notice that as n_a increases ΔF approaches a maximum and then monotonically decreases.   The initial increase of free energy is due to the dominance of the surface free-energy term at low n_a.   This increase is known as the free energy of nucleation and is denoted in the diagram by ΔF^*.   Although the overall reaction results in $\Delta F < 0$, as is required by the laws of thermodynamics, there is always an initial $\Delta F > 0$ due to nucleation.   By convention all aggregates of size less than that at the free-energy maximum (which are unstable relative to re-solution) are called embryos, whereas those of greater size are called crystals. The right-hand side of Fig. 2-3 illustrates that larger crystals are more stable (lower free energy) than smaller crystals.   This explains why larger crystals can grow at the expense of smaller crystals during the aging of a precipitate.

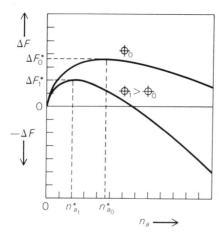

Fig. 2-3 Free energy change ΔF during the formation of a crystal containing n_a ions or atoms.   Values marked with a star * refer to the critical nucleus. (*Adapted from Nielsen, 1964.*) $\oplus = \mathscr{K}T\ln \mathrm{IAP}_0/K$.   Note that n_a^* decreases with increasing $\oplus$.

The size of the critical nucleus (Nielsen, 1964) is derived by differentiating Eq. (75):

$$\frac{d\Delta F}{dn_a} = -\oplus + \sigma\frac{dA}{dn_a} \tag{79}$$

From Eqs. (76) and (79)

$$\frac{d\Delta F}{dn_a} = -\oplus + \sigma v_a\frac{dA}{dV} \tag{80}$$

Now from Eq. (77)

$$\frac{dA}{dV} = \frac{2b}{3V^{\frac{1}{3}}} = \frac{2A}{3V} \tag{81}$$

If the nominal radius r of a crystal is defined as

$$r = 3V/A \tag{82}$$

then

$$\frac{d\Delta F}{dn_a} = -\oplus + \frac{2\sigma v_a}{r} \tag{83}$$

Now if $r = r^*$, the size of the critical radius at the maximum point in Fig. 2-3, the derivative in Eq. (83) is equal to zero. Therefore

$$r^* = \frac{2\sigma v_a}{\oplus} \tag{84}$$

In Fig. 2-4 r^* is plotted versus $\log_{10}(\text{IAP}_0/K)$ or $\oplus/2.3\mathcal{K}T$ for various values of σ. As can be seen the value of r^* increases rapidly at low supersaturation and is considerably increased by an increase in specific surface free energy σ.

Two kinds of nucleation are distinguished: homogeneous and heterogeneous. Homogeneous nucleation results from chance collisions of a sufficient number of dissolved ions to form a crystal embryo of critical size r^* which can then grow by a decrease in free energy. In natural waters homogeneous nucleation is unlikely. Instead heterogeneous nucleation, or nucleation on pre-existing seed crystals, is more likely because of abundant suspended material. In heterogeneous nucleation the free-energy barrier ΔF^* is lowered because of the nucleating seeds. This is due to a lower interfacial surface free energy for the crystal face(s) attached to the seed than for the same face(s) in contact with the precipitating solution.

Because fine crystals, less than one micron, have excess surface free energy, they also are more soluble than larger crystals. For metastable equilibrium between a fine crystal and an associated solution, Eq. (2) is modified to

$$\Delta F^\circ - \tfrac{2}{3}\sigma\overline{A} = -RT\ln K \tag{85}$$

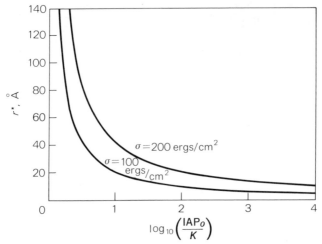

Fig. 2-4 Plot of $\log_{10}(IAP_0/K)$ versus r^* for various values of σ. $T = 25°C$, $v_a = 10^{-22}$ cm^3. (*Adapted from Nielsen, 1964.*)

where $\overline{A}$ is surface area per mole of crystals.   Equation (85) can be rewritten as

$$\Delta F° - \frac{\frac{2}{3}\sigma v B'}{r} = -RT \ln K \tag{86}$$

where r = edge size or radius of crystal

$\quad\quad v$ = molar volume of solid

$\quad\quad B' = Ar/V$ for crystal (for cubes $B' = 6$; for spheres $B' = 3$)

Equation (86) can be used to calculate either the solubility product constant for a given crystal size or the size of crystal in equilibrium with a solution of given concentration.

RATES OF NUCLEATION AND CRYSTAL GROWTH

Nucleation generally follows a rate law (Nielsen, 1964) of the type

$$N = \bar{v} \exp\left[\frac{-4b^3\,\sigma^3\,v_a{}^2}{27\,\oplus^2\,\mathscr{K}\,T}\right] \tag{87}$$

where N = number of nuclei per unit volume per unit time

$\quad\quad \bar{v}$ = "pre-exponential factor," which is on the order of 10^{30}

$\quad\quad \mathscr{K}$ = Boltzmann constant

Note that N increases considerably as σ decreases and as $\oplus$ increases.   For heterogeneous nucleation the substitute term for σ is less than σ^3 and therefore

N is much greater for heteronucleation than for homonucleation. At high degrees of supersaturation (large $\oplus$), the nucleation rate may be so great that most of the excess dissolved mass is precipitated as critical nuclei leaving little material for further growth. As a result the precipitate consists of extremely small crystallites on the order of 10 to 100-Å radius (Fig. 2-4). This explains the common observation that very insoluble substances often precipitate initially as x-ray amorphous "gels."

At lower supersaturation, often a negligible proportion of excess dissolved mass is used to form nuclei. If the nucleation rate is considerably less than the growth rate, then crystals start to grow from initially formed nuclei at about the same time, and the final crystals are all approximately the same size. A typical sigmoid-shaped curve for growth of this type is shown in Fig. 2-5. The nucleation time during which no appreciable change in α_t occurs is known as the induction period. The shape of the sigmoid growth curve can be used to elucidate the type of growth mechanism assuming that the number of crystals remains constant during growth. If nucleation of new crystals during growth is important, then theoretical interpretation of α_t versus t curves becomes more complex.

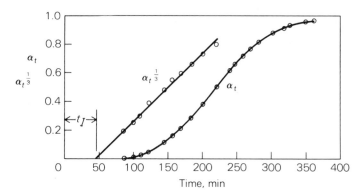

Fig. 2-5 Plot of degree of precipitation, α_t, versus time. α_t is the degree of reaction and is given by the expression

$$\alpha_t = \frac{C_0 - C}{C_0 - C_s}$$

where C = concentration at time t
C_0 = concentration at start of experiment ($t = 0$)
C_s = concentration at equilibrium ($\alpha_t = 1$)
The data are for the crystallization of marcasite on elemental sulfur in saturated H_2S solution at pH = 2.44, $T = 60°C$. The linear plot of $\alpha_t^{\frac{1}{3}}$ versus t is that predicted for a polynuclear growth mechanism. The time t_1, where the α_t plot intercepts the time axis, is the induction period during which nucleation occurs.

Crystal growth is of two basic types: diffusion-controlled growth and surface-reaction-controlled growth.   In diffusion-controlled growth, surface reaction is so fast that ions or atoms are attached to a growing crystal as fast as they can diffuse to its surface.   The rate-limiting step is diffusion and the crystal grows according to the equation (for derivation of this equation see Frank, 1950, or Nielsen, 1961)

$$\frac{dr}{dt} = \frac{Dv(C_\infty - C_S)}{r} \qquad (88)$$

where r = radius (assuming spherical crystals)

D = diffusion coefficient

t = time

v = molar volume of crystal

C_∞ = concentration at a distance from the crystal much greater than its radius, or the bulk concentration out in solution

C_S = concentration at the surface of the crystal assumed to be the equilibrium concentration

In the more general case of precipitation by the reaction of oppositely charged ions the problem is more complex, since Eq. (88) then applies to each ion. Calculation of C_S for a single ion entails equilibrium calculations which are complicated if the ion may dissociate or associate with other ions in solution (a good example is CO_3^{--}).   In the case where neither the cation nor anion of a growing crystal reacts to an appreciable extent with ions in solution, the charge flux to the crystal surface is carried only by the precipitating ions and Eq. (88) may be rewritten, for ions of equal valence (Nielsen, 1964), as:

$$\frac{dr}{dt} = \frac{vD}{r} \left\{ \frac{C_{A^+}^\infty + C_{B^-}^\infty}{2} - \left[K_C + \frac{(C_{A^+}^\infty - C_{B^-}^\infty)^2}{4} \right]^{\frac{1}{2}} \right\} \qquad (88a)$$

where $K_C = (C_{A_S^+} C_{B_S^-})$ the equilibrium product at the surface of the crystal (which at low to moderate salinities is equivalent to K_m)

$C_{A^+}^\infty$; $C_{B^-}^\infty = C_\infty$ for each ion

Equation (88a) can be considerably simplified for two limiting situations:

1. If $|C_{A^+}^\infty - C_{B^-}^\infty| \leqq K_C^{\frac{1}{2}}$:

$$\frac{dr}{dt} \approx \frac{vD}{r} [(ICP)^{\frac{1}{2}} - K_C^{\frac{1}{2}}] \qquad (88b)$$

where ICP $= C_{A^+}^\infty C_{B^-}^\infty$, and at low to moderate salinities is equivalent to IMP.

2. If $C_{A+}^\infty >> C_{B-}^\infty$:

$$\frac{dr}{dt} \approx \frac{vDC_{B-}^\infty}{r} \tag{88c}$$

with the converse being true if $C_{B-}^\infty >> C_{A+}^\infty$.

In surface-reaction-controlled growth the rate of attachment of dissolved ions and atoms to the crystal is so slow that it is the rate-controlling step. As a result the concentration at the surface of the crystal, which represents a steady state between diffusion to the surface and attachment to the crystal, is higher than in the case of diffusion-controlled growth. The slower the surface attachment, the higher is the surface concentration, with the limit for surface concentration being equal to C_∞ where diffusion (and growth) would be infinitely slow.

There are several types of surface-reaction-controlled growth. Using the classification scheme of Nielsen (1964) they are (1) mononuclear growth, (2) polynuclear growth, (3) spiral-dislocation-controlled growth. In mononuclear growth the rate of formation of a single two-dimensional nucleus on a growing face is the rate-governing step. Once the nucleus is formed, spreading outward from it to form a new growth layer is so fast that the layer is completed before an additional surface nucleus is created. In other words each growth layer forms from one surface nucleus. This type of growth should occur only at low supersaturation, where the nucleation rate is low relative to surface spreading. The rate law for mononuclear growth is

$$\frac{dr}{dt} = k_m C^m r^2 \tag{89}$$

where k_m = mononuclear proportionality constant (a function of the nucleation rate constant)

C = concentration in solution

m = integer

Polynuclear growth occurs at higher supersaturations where surface nucleation is sufficiently rapid so that more than one nucleus forms on a crystal face before the face is covered by a new growth layer. Both nucleation and surface spreading are rate-controlling. If the surface spreading is assumed to be diffusion-controlled, the rate law is

$$\frac{dr}{dt} = k_p C^{(m+2)/3} \tag{90}$$

where k_p = polynuclear proportionality constant (a function of the nucleation rate constant and the surface diffusion coefficient).

In spiral-dislocation-controlled growth there is no need for surface nucleation to occur because the surface step of the dislocation acts as a "built-in" nucleus. Because of spiral growth the step is never covered and it acts like a moving nucleus. The constant presence of a surface step enables growth at very low supersaturation, where rates of formation of new surface nuclei are too low for measurable growth by mononuclear or polynuclear mechanisms. At low supersaturation the rate law for spiral-dislocation-controlled growth is

$$\frac{dr}{dt} = k_d (C - C_S)^2 \tag{91}$$

where k_d = dislocation-growth proportionality constant. At higher supersaturation, spiral growth may become so rapid that diffusion becomes the rate-controlling process and surface reaction is no longer important.

Actual growth of a crystal is not restricted to any one mechanism, and a change in mechanism often occurs during growth. Disregarding dislocation-controlled growth, examination of the forms of the other rate laws suggests that in a closed system a crystal may at higher supersaturation start to grow by diffusion, and during growth and consequent drop of supersaturation change successively to polynuclear and mononuclear mechanisms.

RATE OF DISSOLUTION

Dissolution is usually diffusion-controlled. This means that the flux of material from a dissolving surface follows Fick's first law of diffusion, Eq. (63),

$$J = -D \frac{\partial C}{\partial x} \bigg|_x \tag{92}$$

where X refers to distance to the surface relative to a fixed point. If the surface is planar and dissolves uniformly it will retreat at a velocity

$$\frac{dX}{dt} = -Jv \tag{93}$$

where v = molar volume of the dissolving solid. From Eqs. (92) and (93)

$$\frac{dX}{dt} = Dv \frac{\partial C}{\partial x} \bigg|_x \tag{94}$$

If the surface is spherical and dissolves uniformly

$$\frac{dr}{dt} = Dv \frac{\partial C}{\partial \rho} \bigg|_r \tag{95}$$

where r = radius of a spherical body.

The value of $(\partial C/\partial x)_x$ or $(\partial C/\partial \rho)_r$ depends on whether the surrounding water is quiescent or undergoing turbulent mixing. If turbulence is considerable, e.g., as in a surf zone, and the dissolving body is large enough so that it is not carried along completely by the water, a stagnant boundary layer may form around the dissolving body. The thickness of the boundary layer depends upon the rate of stirring, with faster stirring producing a thinner layer. If a boundary layer of l thickness is produced, Eqs. (94) and (95) become

$$\frac{dr}{dt} = \frac{Dv}{l}(C - C_s) \tag{96}$$

$$\frac{dX}{dt} = \frac{Dv}{l}(C - C_s) \tag{97}$$

where C_s = concentration at the surface which is assumed to be the equilibrium solubility

C = uniform concentration in the rapidly mixed solution $(C < C_s)$

If the mass of dissolved material in the water is large compared to the body being dissolved (an open system), C can be treated as a constant and the equations solved for constant l. If, however, C changes as a result of dissolution (a closed system), solution of the differential equations is much more difficult and will not be discussed here. The reader is referred to chapter 10 of Nielsen (1964) for further discussion.

Equation (96) can be rewritten in terms of C rather than r as

$$\frac{\partial C}{\partial t} = \frac{D\bar{A}}{l}(C_s - C) \tag{98}$$

where $\bar{A}$ = total surface area of material undergoing dissolution per unit volume of water.

If the solid undergoing dissolution is present in considerable excess of that needed to reach saturation, $\bar{A}$ (and r) may remain essentially constant. In this case Eq. (98) can be integrated to describe the rate of change of concentration in a closed system as a result of dissolution. It applies only to large bodies in turbulent water which are surrounded by a stagnant boundary layer.

In quiescent water or for small crystals which are carried along by turbulent eddies derivation of the expression for $(\partial C/\partial x)_x$ or $(\partial C/\partial \rho)_r$ is more complex. However, for the spherical case, dissolution is simply the opposite of diffusion-controlled crystal growth, the equation for which is given by Eq. (88). Thus, for dissolution,

$$\frac{dr}{dt} = \frac{vD(C_\infty - C_s)}{r} \tag{88}$$

The only difference between diffusion-controlled growth and dissolution is that in growth $C_\infty > C_s$ (supersaturation) and in dissolution $C_\infty < C_s$ (undersaturation). Integrating Eq. (88) and solving for t,

$$t = \frac{r^2}{2vD(C_s - C_\infty)} \tag{99}$$

Equation (99) gives the time necessary to dissolve a spherical body of radius r by diffusion in quiescent water. It applies to spheres of any size. It also applies to small spheres (<10 microns) which are dissolving in a stirred solution of constant concentration C_∞ (open system). By analogy with Eq. (98) Eq. (88) can be rewritten as

$$\frac{dC}{dt} = \frac{D\bar{A}(C_s - C_\infty)}{r} \tag{100}$$

COLLOID CHEMISTRY

The fundamental theory which is applicable to the colloid chemistry of sediment particles is that of the *electrical double layer*. An electrical double layer is set up at the surface of a solid particle when suspended in an electrolyte solution. If specific surface area is high, which is the case for colloidal particles (<1 micron), then double-layer effects become important. Electrical double layers are of two types:

Type I   Due to imperfections or substitutions within the body of a solid an electrical charge is produced on its surface. The charge is balanced by an excess concentration of ions of opposite charge called *counterions* attracted to the surface from the surrounding solution. The charged surface is referred to as the *fixed-layer* and the balancing counterions as the *mobile layer*. Together they make up the *electrical double layer*.

Type II   Due to specific chemical forces, ions are adsorbed at the surface of a solid causing it to become charged. The adsorbed ions are called *potential-determining ions* and constitute the *fixed layer*. The *mobile layer* is made up of oppositely charged counterions as in type I. Most colloidal substances fall into this category. A diagram illustrating the two types of double layers is shown in Fig. 2-6.

In double layers of type II the charge of the fixed layer may change as the composition of the external solution changes. Following the laws of adsorption the concentration of a given adsorbed ion on a surface decreases as its concentration in solution decreases. In addition, both cations and anions can be adsorbed simultaneously. Thus, as the concentration of a potential-

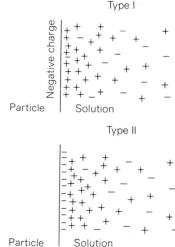

Fig. 2-6 Electrical double layers. Type I is due to an interior lattice charge; type II to the adsorption of potential determining ions. Positive and negative ions are denoted by + and − respectively.

determining cation (which is also a counterion) increases, a point is reached where the net charge on the surface passes from negative to positive. This crossover point is called the *zero point of charge*. In general the zero point of charge does not occur at equal concentrations in solution of potential-determining cations and anions since adsorption is due to specific chemical effects which vary from ion to ion. If the adsorbed ions are H^+ and OH^- or positive and negative species formed from H^+ and OH^- by hydrolysis, the pH at which the zero point of charge is reached is known as the isoelectric point. In double layers of type I the phenomena discussed above are not encountered because the charge on the surface is due to crystallographic causes which are independent of solution composition.

The mobile layer is able to freely exchange counterions with the external solution. As a result an exchange equilibrium is set up with the solution. Most natural sedimentary colloids are negatively charged so that positively charged counterions are exchanged. Such cation exchange can be described for many substances by the simple reaction

$$A_{aq}^+ + BX_{surf} \leftrightarrows B_{aq}^+ + AX_{surf}$$

with

$$K = \frac{a_{B^+} a_{AX_{surf}}}{a_{A^+} a_{BX_{surf}}} \tag{101}$$

For exchange of cations of unlike charge the equilibrium relation is dependent on the type of material undergoing exchange so that no simple mass action type equilibrium constant can be given. In addition a given substance, such as a clay mineral, may have more than one type of exchange site giving rise to more complex expressions, involving different K values, for the overall exchange. The total cation exchange capacity for colloids is usually expressed in terms of milliequivalents per 100 grams dry weight of solid.

Counterions within the mobile layer are affected by two opposing processes. The ions are attracted electrostatically to the oppositely charged fixed layer which causes their concentration to increase relative to the external solution. Increased concentration, however, results in a concentration gradient which causes the counterions to diffuse away from the fixed layer. At equilibrium the forces of diffusion and electrostatic attraction are equal, and the resultant distribution is of the type shown in Fig. 2-7, which is based on simple Gouy theory (Van Olphen, 1963). Ions of the same charge as the fixed layer are repulsed by it but also experience opposing diffusional forces. As a result they too exhibit an equilibrium distribution (see Fig. 2-7). If the fixed layer is negatively charged the latter phenomenon is known as anion exclusion. Strictly speaking, the counterion distribution as described by the simple Gouy theory is incorrect for the region immediately adjacent to the charged surface. However, most of the mobile layer follows a Gouy distribution and more exact modifications of the Gouy theory can be neglected for the purposes of the present discussion. In fact, the mobile layer is commonly referred to as the *Gouy layer*.

As a result of differences in concentration of charged species within the Gouy layer, a gradient in electrical potential arises. The expression relating electrical-potential difference to concentration difference is derived by introducing a new parameter, the electrochemical potential:

$$\mu_{el}^{(+)} = \mu^{(+)} + |Z^{(+)}|\mathscr{F}E \tag{102}$$

$$\mu_{el}^{(-)} = \mu^{(-)} - |Z^{(-)}|\mathscr{F}E \tag{103a}$$

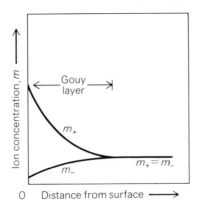

Fig. 2-7 Ion distribution about a negatively charged colloid according to Gouy theory. (*After Van Olphen, 1963.*)

where μ_{el} = electrochemical potential

μ = chemical potential

$|Z|$ = valence charge (absolute value)

(+) and (−) refer to cations and anions, respectively

$\mathscr{F}$ = the Faraday constant

E = electrical potential

The chemical potential μ is the partial molal free energy as defined in Appendix III. It is related to activity by the relation

$$\mu = \mu^\circ + RT \ln a \tag{103b}$$

where $\mu^\circ = \mu$ in the standard state. At equilibrium the electrochemical potential is the same at all points in space. Therefore

$$\mu_I^{(+)} + |Z^{(+)}| \mathscr{F} E_I = \mu_{II}^{(+)} + |Z^{(+)}| \mathscr{F} E_{II} \tag{104}$$

$$\mu_I^{(-)} - |Z^{(-)}| \mathscr{F} E_I = \mu_{II}^{(-)} - |Z^{(-)}| \mathscr{F} E_{II} \tag{105}$$

where I and II refer to two different points.

From Eqs. (103b), (104), and (105)

$$\frac{RT}{|Z^{(+)}| \mathscr{F}} \ln \frac{a_I^{(+)}}{a_{II}^{(+)}} = E_{II} - E_I = \frac{RT}{|Z^{(-)}| \mathscr{F}} \ln \frac{a_{II}^{(-)}}{a_I^{(-)}} \tag{106}$$

Now if E_{ex} refers to a point outside the Gouy layer where there is no excess charge and $\bar{E}$ refers to a space average of E within the Gouy layer, we obtain

$$(E_{ex} - \bar{E}) = \frac{RT}{|Z^{(+)}| \mathscr{F}} \ln \frac{\bar{a}^{(+)}}{a_{ex}^{(+)}} = \frac{RT}{|Z^{(-)}| \mathscr{F}} \ln \frac{a_{ex}^{(-)}}{\bar{a}^{(-)}} \tag{107}$$

where $\bar{a}$ = space average of the activity of either ion within the Gouy layer

a_{ex} = activity outside the Gouy layer

From Eq. (107) one also obtains

$$\left(\frac{\bar{a}^{(+)}}{a_{ex}^{(+)}} \right)^{|Z^{(-)}|} = \left(\frac{\bar{a}^{(-)}}{a_{ex}^{(-)}} \right)^{-|Z^{(+)}|} \tag{108}$$

The potential difference $(E_{ex} - \bar{E})$ is known as the *Donnan potential* and Eq. (108) is called the *Donnan equilibrium*. When $|Z^{(+)}| = |Z^{(-)}|$ Eq. (108) reduces to

$$\bar{a}^{(+)} \bar{a}^{(-)} = a_{ex}^{(+)} a_{ex}^{(-)} \tag{109}$$

REFERENCES

Åkerlöf, G., 1934, The calculation of the composition of an aqueous solution saturated with an arbitrary number of highly soluble strong electrolytes, *Jour. Am. Chem. Soc.*, v. 41, pp. 1439–1443.

Broecker, W. S., and Oversby, V. M., 1970, *Chemical equilibria in the earth*, McGraw-Hill, New York, 336 p.

Denbigh, K., 1963, *The principles of chemical equilibrium*, Cambridge, London, 494 p.

Frank, F. C., 1950, Radially symmetric phase growth controlled by diffusion, *Proc. Roy. Soc. London*, Series A, v. 201, pp. 586–599.

Garrels, R. M., and Christ, C. L., 1965, *Solutions, minerals, and equilibria*, Harper, New York, 450 p.

Harned, H. S., and Owen, B. B., 1958, *The physical chemistry of electrolyte solutions*, Reinhold, New York, 803 p.

Helgeson, H. C., 1964, *Complexing and hydrothermal ore deposition*, Pergamon, New York, 128 p.

Ives, D. J. G., and Janz, G. J., 1961, *Reference electrodes*, Academic, New York, 651 p.

Lewis, G. N., and Randall, M., 1961, *Thermodynamics*, 2d ed., rev. by Pitzer, K. S., and Brewer, L., McGraw-Hill, New York, 723 p.

Nielsen, A. E., 1961, Diffusion controlled growth of a moving sphere. The kinetics of crystal growth in potassium perchlorate precipitation, *Jour. Phys. Chem.*, v. 65, pp. 46–49.

———, 1964, *Kinetics of precipitation*, Macmillan, New York, 151 p.

Robinson, R. A., and Stokes, R. H., 1959, *Electrolyte solutions*, Butterworth, London, 559 p.

Van Olphen, H., 1963, *Clay colloid chemistry*, Interscience, New York, 301 p.

3
Ion Activities
in Natural Waters

To be able to calculate the relative stability of a given mineral in a natural water, a knowledge of ion activity in solution is required. This chapter discusses methods for the calculation of activities from analytical concentrations (molalities) for the three major classes of natural waters: fresh waters, sea water, and supersaline brines.

FRESH WATERS

Most rivers and lakes have an ionic strength less than 0.1 and are thus within the realm of applicability of the Debye-Hückel equation for the calculation of activity coefficients for free ions. Complexing and ion-pair formation must be considered and corrections applied before direct use of the Debye-Hückel expression, especially in the upper ranges of ionic strength (≈ 0.005 to 0.1). At lower ionic strengths ion-pairing can often be neglected. The effect of pressure on activity in general can be neglected because of the relatively shallow depths of natural fresh-water bodies. The effect of temperature on activity

Table 3-1   Composition of world average river water
(Livingstone, 1963)
Calculated activities, a, are for 25°C

Species	ppm	m_T	Calculated a
HCO_3^-	58.3	9.55×10^{-4}	8.90×10^{-4}
SO_4^{--}	11.2	1.17×10^{-4}	0.88×10^{-4}
Cl^-	7.8	2.20×10^{-4}	2.09×10^{-4}
Ca^{++}	15.0	3.75×10^{-4}	2.98×10^{-4}
Mg^{++}	4.1	1.68×10^{-4}	1.35×10^{-4}
Na^+	6.3	2.74×10^{-4}	2.60×10^{-4}
K^+	2.3	0.59×10^{-4}	0.56×10^{-4}

is accounted for by choosing the appropriate temperature-dependent parameters, A and B in the Debye-Hückel equation, and if needed also by calculating the K values for ion-complex formation from K values at 25°C using Eq. (5) of Chapter 2. Unfortunately, $\Delta h°$ values for ion association are frequently not known but in some cases approximations based on analogous equilibria can be made.

As an example of the calculation of ion activities in natural fresh waters, the world average for river water (Livingstone, 1963) is chosen. The major ion composition of average river water is given in Table 3-1. The ionic strength, before correcting for possible ion-pairing, calculated directly from the data of Table 3-1, is

$$I = \tfrac{1}{2}(4m_{T_{SO_4}--} + 4m_{T_{Ca}++} + 4m_{T_{Mg}++} + m_{T_{HCO_3}-} + m_{T_{K}+} + m_{T_{Cl}-} + m_{T_{Na}+})$$
$$= 0.0021 \tag{1}$$

This value is low enough so that inorganic complexing should not be important. This is demonstrated by calculation. The strongest ion pairs are those between doubly charged ions; therefore, consider $MgSO_4^0$. From Table 3-2 we obtain

$$\frac{a_{Mg^{++}} a_{SO_4^-}}{a_{MgSO_4^0}} = 10^{-2.36} \tag{2}$$

As a first approximation it is assumed that $m_{SO_4^{--}} \approx m_{T_{SO_4}}--$ and that $a_{SO_4^{--}} = \gamma_{SO_4^{--}} m_{T_{SO_4}}--$ (where $m_{T_{SO_4}}-- = m_{SO_{4T}}-- =$ total or analytical molality). Calculation of $\gamma_{SO_4^{--}}$ from the ionic strength given above and the Debye–Hückel equation gives $a_{SO_4^{--}} \approx 10^{-4.02}$. Substituting this value in Eq. (2),

$$\frac{a_{Mg^{++}}}{a_{MgSO_4^0}} \approx 10^{1.66} \tag{3}$$

Table 3-2 Ion-pair dissociation constants for the major ions of natural waters at 25°C, 1-atm pressure
$K = a_A a_B / a_{AB}$; $pK = -\log K$; values of pK are derived from ΔF_f° values in Appendix I

Cations	HCO_3^-	CO_3^{--}	SO_4^{--}	Cl^-
		pK		
K^+	<0	<0	0.96	<<0
Na^+	−0.25	1.27	1.06	<<0
Mg^{++}	1.16	2.4	2.36	<<0
Ca^{++}	1.26	3.2	2.31	<<0

If $\gamma_{MgSO_4^0} = 1$, which is accurate for low ionic strength, substitution of $a_{Mg^{++}} = \gamma_{Mg^{++}} m_{Mg^{++}}$ in Eq. (3) yields

$$\frac{m_{MgSO_4^0}}{m_{Mg^{++}}} \approx 0.018 \tag{4}$$

The concentration of $MgSO_4^0$ is therefore approximately 2 percent of the total dissolved magnesium. By analogous calculation the concentration of $CaSO_4^0$ is also found to be about 2 percent of the total dissolved calcium. The concentrations of $CaHCO_3^+$ and $MgHCO_3^+$ by similar calculation amount to about 1 percent of the total dissolved concentrations of Ca^{++} and Mg^{++}. The activity of sulfate is affected by pairing with all four major cations. Because of the low concentrations and weak interactions with Na^+ and K^+, ion pairs with these ions can be neglected. The remaining mass-balance equation is

$$m_{SO_4^{--}} = m_{T_{SO_4^{--}}} - (m_{CaSO_4^0} + m_{MgSO_4^0}) \tag{5}$$
$$= 1.17 \times 10^{-4} - m_{CaSO_4^0} - m_{MgSO_4^0}$$

If $m_{CaSO_4^0} \approx 0.02 \, m_{T_{Ca^{++}}}$ and $m_{MgSO_4^0} \approx 0.02 \, m_{T_{Mg^{++}}}$,

$$m_{SO_4^{--}} \approx 1.06 \times 10^{-4} \tag{6}$$

Back substitution of this value into Eq. (2) leads to the corrected values

$$\frac{m_{MgSO_4^0}}{m_{Mg^{++}}} = 0.016 \tag{7}$$

and

$$m_{SO_4^{++}} = 1.08 \times 10^{-4} \tag{8}$$

Further back substitution and recalculation result in negligible correction.

As a result of the above calculations it is found that $m_{Ca^{++}} = 0.97$ $m_{Ca_T^{++}}$, $m_{Mg^{++}} = 0.97$ $m_{Mg_T^{++}}$, $m_{SO_4^{--}} = 0.92$ $m_{SO_{4_T}^{--}}$, and $m_{HCO_3^-} = 0.98$ $m_{HCO_{3_T}^-}$. All other ions of Table 3-1 are complexed by far less than 1 percent of their total amounts. Thus, as predicted earlier, the effect of inorganic complexing is small at an ionic strength of 0.0021 and the composition given in Table 3-1. For crude geochemical calculations (± 10 percent accuracy) complexing could have been ignored. If complexing had been more important, then a recalculation of the ionic strength due to a redistribution of charge and formation of new species would have been necessary. Thus at higher ionic strengths ion-activity calculations become considerably more complicated. This is the case for moderately saline rivers and lakes and brackish marginal marine waters.

SEA WATER

Calculation of ionic activities in sea water is complicated. The average ionic strength is about 0.7 and complexing is very important for some species. Depths in the ocean are great enough so that the presssure effect on ion-association constants must be considered. Temperatures range from less than 0°C to about 40°C. A simplifying factor, however, is that more than 99 percent of sea water falls within a narrow range of salinity and exhibits a constant ratio between the concentrations of the major ions (Sverdrup et al., 1942). Only near sources of fresh water such as river mouths and melting glaciers, or in relatively isolated arms of the sea where evaporation or anaerobiosis may greatly affect the water, do the salinity and composition diverge greatly from the average. Thus, the compositional homogeneity of sea water enables calculation of ion activities for an average salinity which is characteristic for the whole ocean. This average salinity corresponds to an ionic strength of ≈ 0.7.

The effect of temperature, and especially pressure, on activity coefficients and ion-association equilibrium constants is in general poorly known. Therefore, calculation of ion activities is presented here only where data are available, that is 25°C and 1-atm total pressure. This calculation is a reasonably accurate approximation for shallow sea water at low latitudes and midlatitudes during warmer seasons.

The calculation of ion activities for sea water at $I = 0.7$, $T = 25$°C, $P = 1$ atm has been performed by Garrels and Thompson (1962) for the ions Cl^-, Na^+, Mg^{++}, SO_4^{--}, K^+, Ca^{++}, HCO_3^-, and CO_3^{--}. Their approach includes the mean-salt method to obtain γ for free ions and the assumption that only simple ion pairs constitute all complexes affecting activity. Another assumption which has to be made in the absence of concrete data is that γ for all univalent ion pairs (divalent pairs are not encountered) is the same as γ

Table 3-3   Individual ion-activity coefficients γ used to calculate the distribution of dissolved species in sea water, according to the model of Garrels and Thompson (1962)

Values for Cl^-, Na^+, K^+, Ca^{++}, and Mg^{++} are recalculated here using data from Robinson and Stokes (1959)

Species	γ	Method
Cl^-	0.63	$= \gamma_{\pm KCl}$
Na^+	0.71	$= \gamma^2_{\pm NaCl}\,\gamma_{\pm KCl}$
K^+	0.63	$= \gamma_{\pm KCl}$
Ca^{++}	0.26	$= \gamma^3_{\pm CaCl_2}/\gamma^2_{\pm KCl}$
Mg^{++}	0.29	$= \gamma^3_{\pm MgCl_2}/\gamma^2_{\pm KCl}$
SO_4^{--}	0.17	$= 1.1\ \gamma_{Debye-Hückel}$ *
CO_3^{--}	0.20	$= \gamma^3_{\pm K_2CO_3}/\gamma^2_{\pm KCl}$
HCO_3^-	0.68	$= \gamma^2_{\pm KHCO_3}/\gamma_{\pm KCl}$
$NaCO_3^-$	0.68	$= \gamma_{HCO_3-}$
$NaSO_4^-$	0.68	$= \gamma_{HCO_3-}$
KSO_4^-	0.68	$= \gamma_{HCO_3-}$
$MgHCO_3^+$	0.68	$= \gamma_{HCO_3-}$
$CaHCO_3^+$	0.68	$= \gamma_{HCO_3-}$
$NaHCO_3^0$	1.13	$= \gamma_{H_2CO_3}$
$MgSO_4^0$	1.13	$= \gamma_{H_2CO_3}$
$CaSO_4^0$	1.13	$= \gamma_{H_2CO_3}$
$MgCO_3^0$	1.13	$= \gamma_{H_2CO_3}$
$CaCO_3^0$	1.13	$= \gamma_{H_2CO_3}$

* Based on the extension of the Debye-Hückel equation for ions showing no ion-pairing (M. Lafon, personal communication).

calculated for $HCO_{3\,aq}^-$ by the mean-salt method.   It is also assumed that γ for neutral dissolved species is the same as that for H_2CO_3 in 0.7 m NaCl solution.   A summary of γ values is presented in Table 3-3.

To calculate the concentrations of each free ion and ion pair, seven mass-balance equations and 10 ion-association equilibrium equations are necessary. Chloride ion is neglected since it does not form ion pairs at $I = 0.7$.   Values of pK for the 10 possible ion pairs between the ions under consideration have been presented already in Table 3-2.   The seven mass-balance equations are

$$m_{K_T^+} = m_{K^+} + m_{KSO_4^-} \qquad\qquad (9)$$

$$m_{Na_T^+} = m_{Na^+} + m_{NaSO_4^-} + m_{NaCO_3^-} + m_{NaHCO_3^0} \qquad\qquad (10)$$

$$m_{Mg_T^{++}} = m_{Mg^{++}} + m_{MgSO_4^0} + m_{MgCO_3^0} + m_{MgHCO_3^+} \tag{11}$$

$$m_{Ca_T^{++}} = m_{Ca^{++}} + m_{CaSO_4^0} + m_{CaCO_3^0} + m_{CaHCO_3^+} \tag{12}$$

$$m_{SO_{4_T}^{--}} = m_{SO_4^{--}} + m_{KSO_4^-} + m_{NaSO_4^-} + m_{MgSO_4^0} + m_{CaSO_4^0} \tag{13}$$

$$m_{HCO_{3_T}^-} = m_{HCO_3^-} + m_{NaHCO_3^0} + m_{MgHCO_3^+} + m_{CaHCO_3^+} \tag{14}$$

$$m_{CO_{3_T}^{--}} = m_{CO_3^{--}} + m_{NaCO_3^-} + m_{MgCO_3^0} + m_{CaCO_3^0} \tag{15}$$

By substituting γm for a in each of the 10 equilibrium expressions, and by using the γ values of Table 3-3, and the measured values of m_T for average sea water (see Table 3-4) one obtains 17 equations in 17 unknowns, which enables calculation of the molality of each species. Rigorous solution of this problem necessitates a digital computer. However, a little chemical intuition enables solution by pencil, paper, and slide rule in a few hours and this latter technique is briefly described here. As a first approximation it is assumed that

$$m_{K^+} = m_{K_T^+} \qquad m_{Na^+} = m_{Na_T^+}$$

$$m_{Mg^{++}} = m_{Mg_T^{++}} \qquad m_{Ca^{++}} = m_{Ca_T^{++}}$$

This is reasonable since the principal anion balancing these cations is Cl^-, which does not form ion pairs. Substitution of these values for free cations into the mass balance and equilibrium equations for SO_4^{--} enables the solution of

Table 3-4   Distribution of major dissolved species in sea water,
$I = 0.7$, $T = 25°C$, $P = 1$ atm, calculated according to the model of
Garrels and Thompson (1962)
Values of m_T are from Turekian (1968)

Ion	$m_T \times 10^3$	Free ion, %	SO_4^{--} ion pair, %	HCO_3^- ion pair, %	CO_3^{--} ion pair, %
Na⁺	475	99	1	—	—
K⁺	10.0	99	1	—	—
Mg⁺⁺	54.0	87	11	1	0.3
Ca⁺⁺	10.4	91	8	1	0.2

			Ca ion pair, %	Mg ion pair, %	Na ion pair, %	K ion pair, %
SO₄⁻⁻	28.4	40	3	19	38	0.5
HCO₃⁻	1.8	74	3	15	9	—
CO₃⁻⁻	0.27	10.5	7	63.5	19	—
Cl⁻	550	100	—	—	—	—

five equations for the five unknowns, $m_{SO_4^{--}}$, $m_{KSO_4^-}$, $m_{NaSO_4^-}$, $m_{MgSO_4^0}$, and $m_{CaSO_4^0}$. Likewise, substitution into the mass-balance and equilibrium equations for HCO_3^- and CO_3^{--} (four equations in each case) enables solution for all remaining species. After this first round of solutions it is seen that only relatively small fractions of Ca_T^{++} and Mg_T^{++} are complexed and that complexes make up an insignificant proportion of Na_T^+ and K_T^+. Thus the original assumptions were reasonably valid. After correcting for the fractions of Ca_T^{++} and Mg_T^{++} complexed by SO_4^{--}, HCO_3^-, and CO_3^{--}, a recalculation using the same method as above is performed. The second calculation is found to be sufficiently accurate as demonstrated by one additional recalculation.

Results are given in Table 3-4, with average values for total measured concentrations. Note that complexing mainly affects the anions, SO_4^{--}, HCO_3^{--}, and CO_3^{--} with the most important complexing ions being Na^+ and Mg^{++}. Neglect of ion-pair formation in the case of CO_3^{--} ion would result in an erroneous activity ten times too high. Divalent anions of lower concentration which have not been considered, such as HPO_4^{--}, and which form strong ion pairs should also show a very small percent of the total that is free ion. Activities of the free ions, using the activity coefficients of Table 3-3 are given in Table 3-5.

Checks on the Garrels and Thompson model have been provided by electrode measurements and solubility studies. Comparison of calculated values and those derived from measurements in terms of γ_T are shown in Table 3-6. Considering the crude approximations of the model good agreement is demonstrated for all ions. This indicates that the Garrels and Thompson approach, for geochemical purposes, is valid at $I = 0.7$.

Table 3-5   Activities of major ions in sea water at $I = 0.7$, $T = 25°C$, $P = 1$ atm, calculated by the Garrels and Thompson model

Species	$m_T \times 10^3$	$a \times 10^3$	$-\log a$
Cl^-	550	346	0.46
Na^+	475	332	0.48
Mg^{++}	54	13.5	1.87
SO_4^{--}	28.4	1.94	2.71
Ca^{++}	10.4	2.39	2.62
K^+	10.0	6.2	2.21
HCO_3^-	1.8	0.92	3.04
CO_3^{--}	0.27	0.0057	5.25

Table 3-6   Comparison of γ_T values for the Garrels and
Thompson (G and T) model for sea water with measured
values $\gamma_T = \gamma m/m_T$

Species	γ_T(G and T)	γ_T(meas.)	Method and reference
Na^+	0.70	0.70	Na-glass electrode (Garrels, 1967)
K^+	0.62	0.60	$K^+ + Na^+$ glass electrodes (Garrels and Thompson, 1962)
Ca^{++}	0.23	0.21	Ca^{++}-membrane electrode (Thompson and Ross, 1966)
Ca^{++}	0.23	0.20	pH and $CaCO_3$ saturation (Berner, 1965)
Mg^{++}	0.25	0.26	Mg^{++}-membrane electrode (Thompson, 1966)
Mg^{++}	0.25	0.26	pH and $Mg(OH)_2$ saturation (Pytkowicz and Gates, 1968)
HCO_3^-	0.51	0.55	pH and H_2CO_3 equilibria (Berner, 1965)
CO_3^{--}	0.021	0.021	pH and H_2CO_3 equilibria (Berner, 1965)
SO_4^{--}	0.068	0.071	Gypsum solubility in sea water calculated from data of Posnjak (1940)

SUPERSALINE BRINES

No a priori theoretical method is available for the calculation of individual ion activities in highly concentrated natural brines.   The type of model used for average sea water may possibly be used for somewhat more concentrated waters (Truesdell and Jones, 1969), but salt solutions with ionic strengths greater than ~1.0 may necessitate more sophisticated models.   At high ionic strengths the mean-salt method must be corrected through the use of Harned's rule (see Chapter 2).   Another important factor is the possibility of complex species made up of more than 2 ions, e.g., ion triplets.   Formation of ion pairs and triplets apparently is so extensive in highly saline sulfate and carbonate brines that the true ionic strength may be less than half the value calculated from total molalities.   The assumptions made for sea water that γ for singly charged ion pairs is the same as that for $HCO_{3\,aq}^-$ and that γ for all neutral species is the same as that for H_2CO_3 can also lead to considerable error at high ionic strength.   As an example of the latter in 2 N NaCl solution ($I = 2.0$), $\gamma_{H_2\,aq} = 1.5$, whereas $\gamma_{N_2\,aq} = 2.7$.

The best method for determining ion activities in highly saline waters is through the use of electrodes.   Even here several complications arise.   One is that of liquid-junction potentials.   If the ion-sensitive electrode plus reference electrode are standardized in a solution which is dilute enough so that an individual ion activity can be calculated via Debye-Hückel or other theories, the liquid-junction potential in such a dilute solution should be considerably different than that in a highly saline brine (Ives and Janz, 1961). If only relative cation activities are desired, chloride solutions of cations at the same concentration as in the natural brines can be used as standard solutions in order to minimize liquid-junction potential errors. The best way to avoid liquid-junction potential problems is to use a combination of cation- and anion-sensitive electrodes. In this manner ion-activity products or ion-activity ratios can be directly measured. Since no reference electrode of fixed potential is used, there is no liquid-junction potential. A knowledge of the product of ion activities in natural waters is often sought in order to test the waters for saturation with respect to certain minerals. Electrode measurement of IAP using combined cation- and anion-sensitive electrodes can achieve this end directly. An example is the use of sodium glass and Ag^0–$AgCl$ electrodes to measure the IAP for NaCl in brines of the Dead Sea (Lerman and Shatkay, 1968).

Another problem pertaining to the use of electrodes in hypersaline brines is the purely technological one of nonselectivity of most electrodes. Saline waters may contain concentrations of interfering species high enough to affect the potentials of electrodes which otherwise would give accurate results for the ions which each electrode is purported to measure. It is hoped that the problems of nonselective interference can be reduced by advances in electrode technology. The glass electrodes used for pH measurement have already been perfected to a level where the activity of H_{aq}^+ can be measured without interference by Na_{aq}^+ at ionic ratios of Na^+/H^+ of 10^{13}.

If electrodes were perfected to the point where they could be used to measure all major ions of sea water without interference error, then a different approach to the determination of ion activities in sea water and all other more saline solutions could be used. The method consists of using cation plus anion electrodes together in different combinations to determine various ion-activity products. For example the product a_{Na^+} times a_{Cl^-} could be calculated from the potential difference between sodium plus chloride electrodes in the natural water and in a NaCl solution where the $\gamma_\pm$ value is known. Likewise IAP for Na_2SO_4, $MgSO_4$, KCl, $CaCl_2$, $NaHCO_3$, and Na_2CO_3 could be determined with different electrodes by comparison of potentials in the natural water with those in solutions of the pure salt where $\gamma_\pm$ is known. In the pure salt solutions

$$\text{IAP} = a_A{}^x a_B{}^y = \gamma_{T_A}{}^x \gamma_{T_B}{}^y m_{T_A}{}^x m_{T_B}{}^y \tag{16}$$

or

$$\text{IAP} = \gamma_\pm{}^{x+y} \text{IMP} \tag{17}$$

and

$$\ln{(IAP_{sample})} = \ln{(IAP_{std})} + \frac{Z\mathscr{F}}{RT}[E_{sample} - E_{std}] \tag{18}$$

In no case are there any liquid-junction potentials to be concerned with so that standard solutions need not be of the same ionic strength as the natural waters. The only precaution is to be sure that the electrode pairs have the proper voltage response $RT/Z\mathscr{F}$ which can be checked by using several different concentrations of the pure salt solution.   If not, the measured voltage response can be still used as a purely empirical calibration curve.

Once IAP values which include all ions of interest are measured, then IAP values for combinations of ions not measured can be directly calculated. This is especially useful for ion products of relatively insoluble salts for which there are no available standard solutions.   For example, if IAP_{CaCO_3} were desired it could be calculated by

$$IAP_{CaCO_3} = \frac{IAP_{CaCl_2} \, IAP_{Na_2CO_3}}{(IAP_{NaCl})^2} \tag{19}$$

For average sea water a simple way of tabulating data would be to *arbitrarily* assume $\gamma_{K_T^+} = \gamma_{Cl_T^-}$.   From this, operational values of γ_T for all other ions could be calculated as follows:

$$IAP_{KCl_{meas}} = IMP_{KCl} \gamma_{K_T^+} \gamma_{Cl_T^-} \tag{20}$$

Since

$$\gamma_{K_T^+} = \gamma_{Cl_T^-}$$

therefore

$$\gamma_{K_T^+} = \left(\frac{IAP_{KCl}}{IMP_{KCl}}\right)^{\frac{1}{2}} = \gamma_{Cl_T^-} \tag{21}$$

Likewise,

$$IAP_{CaCl_2} = IMP_{CaCl_2} \, \gamma_{Ca_T^{++}} \gamma^2_{Cl_T^-} \tag{22}$$

From Eqs. (21) and (22),

$$\gamma_{Ca_T^{++}} = \frac{IAP_{CaCl_2} \, IMP_{KCl}}{IMP_{CaCl_2} \, IAP_{KCl}} \tag{23}$$

and so on for the other ions.

Ion-activity ratios as well as ion-activity products could be determined by the electrode method.   The procedure is to measure IAP for two salts with the same anion (or cation) and then take the ratio of IAP.   For example, suppose the value of $a_{Ca^{++}}/a_{Mg^{++}}$ is sought.   If IAP for $MgCl_2$ and IAP for $CaCl_2$ were measured, then

$$\frac{a_{Ca^{++}}}{a_{Mg^{++}}} = \frac{IAP_{CaCl_2}}{IAP_{MgCl_2}} \tag{24}$$

Although the electrode method gives no insight into the theoretical basis for variations in ion activity (γ_T is a purely empirical parameter), it does provide a direct answer to the major question of what is the IAP or ion-activity ratio in any given water. Since $\gamma_\pm$ values are generally tabulated above $I = 0.1$, the electrode method could be used for waters of higher ionic strength than 0.1 and the Debye-Hückel method (with or without correction for ion pairing) for ionic strength below 0.1.

In the absence of suitable electrodes IAP in hypersaline brines can under some circumstances be calculated via Harned's rule (see Chapter 2), which is

$$\log \gamma_{\pm_1} = \log \gamma_{\pm_{1(0)}} + \sum \alpha_{1,i} I_i \qquad (25)$$

where values of $\gamma_\pm$ refer to a total ionic strength, $I_T = I_1 + \sum I_i$, at which $\alpha_{1,i}$ is a constant independent of composition. The symbol 1(0) refers to a pure solution of salt 1 and all ionic strengths are calculated from total molalities m_T. Lanier (1965), using sodium glass electrodes, has shown that $\gamma_\pm$ for NaCl obeys Harned's rule in a variety of two-salt solutions over a wide range of total ionic strength. His measured values of α for NaCl–MgCl$_2$, NaCl–CaCl$_2$, and NaCl–Na$_2$SO$_4$ are plotted in Fig. 3-1. Åkerlöf (1934) has shown that α values obtained in two-salt mixtures can also be used in multicomponent concentrated electrolyte solutions, as described by Eq. (25).

Additional values of α for NaCl and other salts have been obtained by Åkerlöf (1934, 1937) using solubility measurements made at very high total ionic strengths (6 to 13). The problem with his method is that among other things he assumes that α is not a function of I_T and his values do not agree with

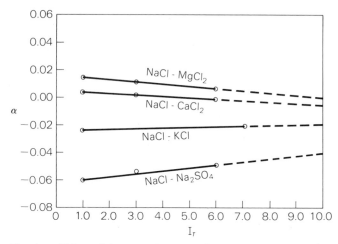

Fig. 3-1 Values of the Harned's rule coefficient α for $\gamma_{\pm NaCl}$ in various solutions at total ionic strength, I_T and $T = 25°C$. [*Data for MgCl$_2$, CaCl$_2$, and Na$_2$SO$_4$ from Lanier (1965). Data for KCl from Åkerlöf (1934) and Robinson and Stokes (1959).*]

extrapolations of Lanier's data in Fig. 3-1. As a result, only Åkerlöf's data for NaCl–KCl which checks with later work (Robinson and Stokes, 1959) are plotted. Values of α for salts other than NaCl based on solubility measurements can be found in Åkerlöf (1934, 1937) and Lerman (1967), but caution should be applied in their usage until further checks can be made.

Once $\gamma_\pm$ values have been calculated, IAP is obtained from Eqs. (17) and (25):

$$\log \text{IAP} = (x + y)\,(\log \gamma_{\pm 1(0)} + \sum \alpha_{1,\,i} I_i) + \log \text{IMP} \tag{26}$$

Application of Eq. (26) to natural waters has been done by Lerman (1967) who calculated IAP_{NaCl} in the bottom waters of the Dead Sea using α values derived from high ionic-strength solubility measurements. He found that IAP_{NaCl} is close to K for halite. Since halite is an abundant constituent in the Dead Sea bottom sediments, Eq. (26) appears to be valid for Dead Sea water which is basically a $MgCl_2$–NaCl–$CaCl_2$ solution.* Use of Harned's rule in calculating the degree of evaporation at which sea water becomes saturated with halite is presented in Chapter 5.

REFERENCES

Åkerlöf, G., 1934, The calculation of the composition of an aqueous solution saturated with an arbitrary number of highly soluble strong electrolytes, *Jour. Am. Chem. Soc.*, v. 56, pp. 1439–1443.
———, 1937, A study of the composition of the liquid phase in aqueous systems containing strong electrolytes of higher valence types as solid phases, *Jour. Phys. Chem.*, v. 41, pp. 1053–1076.
Berner, R. A., 1965, Activity coefficients of bicarbonate, carbonate, and calcium ions in sea water, *Geochim. et Cosmochim. Acta*, v. 29, pp. 947–965.
Garrels, R. M., 1967, Ion-sensitive electrodes and individual ion activity coefficients, in *Glass electrodes for hydrogen and other cations*, Eisenman, G. D. (ed.), Marcel Dekker, New York, pp. 344–361.
———, and Thompson, M. E., 1962, A chemical model for sea water at 25°C and one atmosphere total pressure, *Am. Jour. Sci.*, v. 260, pp. 57–66.
Ives, D. J. G., and Janz, G. J., 1961, *Reference electrodes*, Academic, New York, 651 p.
Lanier, R. D., 1965, Activity coefficients of sodium chloride in aqueous three component solutions by cation-sensitive glass electrodes, *Jour. Phys. Chem.*, v. 69, pp. 3992–3998.
Lerman, A., 1967, Model of chemical evolution of a chloride lake—the Dead Sea, *Geochim. et Cosmochim. Acta*, v. 31, pp. 2309–2330.
———, and Shatkay, A., 1968, Dead Sea brines: Degree of halite saturation by electrode measurements, *Earth and Planet. Sci. Letters*, v. 5, pp. 63–66.
Livingstone, D. A., 1963, Chemical composition of rivers and lakes, *U.S. Geol. Survey Prof. Paper 440-G*, 61 p.

* The coefficients for NaCl–$MgCl_2$ and NaCl–$CaCl_2$ used by Lerman are probably too high since they do not agree with predictions from Fig. 3-1. This is confirmed by later work by Lerman and Shatkay (1968), where they show that IAP calculated for Dead Sea water using the same α coefficients is 15 to 20 percent higher than the value measured with electrodes.

Posnjak, E., 1940, Deposition of calcium sulfate from sea water, *Am. Jour. Sci.*, v. 238, pp. 559–568.

Pytkowicz, R. M., and Gates, R., 1968, Magnesium sulfate interactions in sea water from solubility measurements, *Science*, v. 161, pp. 690–691.

Robinson, R. A., and Stokes, R. H., 1959, *Electrolyte solutions*, Butterworth, London, 559 p.

Sverdrup, H. U., Johnson, M. W., and Fleming, R. H., 1942, *The oceans*, Prentice-Hall, Englewood Cliffs, N.J., 1087 p.

Thompson, M. E., 1966, Magnesium in sea water: An electrode measurement, *Science*, v. 153, pp. 866–867.

Thompson, M. E., and Ross, J. W., 1966, Calcium in sea water by electrode measurement, Science, v. 154, pp. 1643–1644.

Truesdell, A. H., and Jones, B. F., 1969, Ion association in natural brines, *Chemical Geology*, v. 4, pp. 51–62.

Turekian, K. K., 1968, The oceans, streams, and atmosphere, in *Handbook of geochemistry*, v. I, Springer-Verlag, Berlin, pp. 297–323.

4
Calcium Carbonate Chemistry in Surface Waters

In this chapter the state of saturation, factors affecting this state, and the solution and precipitation of calcium carbonate will be discussed. Only overlying waters are treated, the problem of sediment pore waters being covered under separate chapters on diagenesis. Because of its quantitative importance most of the discussion is confined to sea water.

STATE OF SATURATION

METHODS OF CALCULATION

The state of saturation of a given water with respect to $CaCO_3$ can be given by the ratio IAP/K, where K is the equilibrium constant for the reaction

$$CaCO_3 \leftrightarrows Ca_{aq}^{++} + CO_{3\,aq}^{--}$$

The value of K for this reaction can be calculated from the ΔF_f° data of Appendix I and Eq. (2) of Chapter 2. Values of IAP (ion-activity product) are determined by the calculation of $a_{Ca^{++}}$ using the methods of the previous chapters, and $a_{CO_3^{--}}$ using carbonate equilibrium calculations and any two of the four measurements A_C, ΣCO_2, pH, and P_{CO2}. The parameter A_C is the carbonate alkalinity and is defined as

$$A_C = m_{T_{HCO_3^-}} + 2m_{T_{CO_3^{--}}} \tag{1}$$

It is derived from the titration alkalinity A_T, which is the number of equivalents of hydrogen ion (per kilogram of solution) necessary to convert all anions of weak acids to their respective acids. In many waters A_C and A_T are essentially identical. The function ΣCO_2 is defined as

$$\Sigma CO_2 = m_{CO_2} + m_{H_2CO_3} + m_{T_{HCO_3^-}} + m_{T_{CO_3^{--}}} \tag{2}$$

It represents the total concentration of all species of dissolved oxidized inorganic carbon. It can be determined by conversion of all species, via acidification and boiling, to CO_2 gas. The symbols pH and P_{CO_2} refer respectively to the negative logarithm of the activity of hydrogen ion and the partial pressure of carbon dioxide gas.

The procedures for calculating $a_{CO_3^{--}}$ make use of any pair combination of the parameters pH, P_{CO_2}, A_C, and ΣCO_2. The pertinent equilibria relating these measurements to $a_{CO_3^{--}}$ are

$$CO_{2\,gas} + H_2O_{liq} \leftrightharpoons H_2CO_{3\,aq}$$

$$K_0 = \frac{a_{H_2CO_3}}{P_{CO_2} a_{H_2O}} \tag{3}$$

$$H_2CO_{3\,aq} \leftrightharpoons H_{aq}^+ + HCO_{3\,aq}^-$$

$$K_I = \frac{a_{H^+} a_{HCO_3^-}}{a_{H_2CO_3}} \tag{4}$$

$$HCO_{3\,aq}^- \leftrightharpoons H_{aq}^+ + CO_{3\,aq}^{--}$$

$$K_{II} = \frac{a_{H^+} a_{CO_3^{--}}}{a_{HCO_3^-}} \tag{5}$$

For the sake of simplicity the parameter $m_{H_2CO_3}$ is assumed to include m_{CO_2} and represents all neutral dissolved carbon dioxide. To obtain $a_{CO_3^{--}}$ from pH and P_{CO_2} Eqs. (3), (4), and (5) are combined:

$$CO_{2\,gas} + H_2O_{liq} \leftrightharpoons 2H_{aq}^+ + CO_{3\,aq}^{--}$$

$$K = K_0 K_I K_{II} = \frac{a_{H^+}^2 a_{CO_3^{--}}}{P_{CO_2} a_{H_2O}} \tag{6}$$

For most waters discussed in this chapter $a_{H_2O} \approx 1$ so that it can be neglected in calculations. Thus

$$a_{CO_3^{--}} = \frac{K_0 K_I K_{II} P_{CO_2}}{a_{H^+}^2} \tag{7}$$

To obtain $a_{CO_3^{--}}$ from pH and A_C (the most common procedure), use is made of Eqs. (1) and (5).   Rewriting Eq. (1) in terms of a and γ_T:

$$A_C = \frac{a_{HCO_3^-}}{\gamma_{T_{HCO_3^-}}} + \frac{2a_{CO_3^{--}}}{\gamma_{T_{CO_3^{--}}}} \tag{8}$$

Substitution of (5) and solution for $a_{CO_3^{--}}$ yields

$$a_{CO_3^{--}} = \frac{A_C}{\left(\dfrac{a_{H^+}}{K_{II}\,\gamma_{T_{HCO_3^-}}} + \dfrac{2}{\gamma_{T_{CO_3^{--}}}} \right)} \tag{9}$$

For pH values of 7.5 or less in sea water and 8.5 or less in fresh water $m_{T_{HCO_3^-}} \gg m_{T_{CO_3^{--}}}$ so that A_C can be set equal to $m_{T_{HCO_3^-}}$.   In this case Eq. (9) simplifies to

$$a_{CO_3^{--}} = \frac{A_C\, K_{II}\, \gamma_{T_{HCO_3^-}}}{a_{H^+}} \tag{10}$$

The activity of CO_3^{--} is calculated from pH and ΣCO_2 by substitution of Eqs. (4) and (5) in Eq. (2).   [m_{CO_2} is dropped from Eq. (2) as it is included in $m_{H_2CO_3}$.]   The resulting expression is

$$a_{CO_3^{--}} = \frac{\Sigma CO_2}{\dfrac{a_{H^+}^2}{K_I\,K_{II}\,\gamma_{H_2CO_3}} + \dfrac{a_{H^+}}{K_{II}\,\gamma_{T_{HCO_3^-}}} + \dfrac{1}{\gamma_{T_{CO_3^{--}}}}} \tag{11}$$

To obtain $a_{CO_3^{--}}$ from P_{CO_2} and A_C, substitution of Eqs. (3), (4), and (5) in (1) yields

$$A_C = \left(\frac{K_0\,K_I\,a_{CO_3^{--}}\,P_{CO_2}}{K_{II}\,\gamma_{T_{HCO_3^-}}^2} \right)^{\frac{1}{2}} + \frac{2a_{CO_3^{--}}}{\gamma_{T_{CO_3^{--}}}} \tag{12}$$

Similarly for P_{CO_2} and ΣCO_2 substitution of Eqs. (3), (4), and (5) in (2) yields

$$\Sigma CO_2 = \frac{K_0\,P_{CO_2}}{\gamma_{H_2CO_3}} + \left(\frac{K_0\,K_I\,a_{CO_3^{--}}\,P_{CO_2}}{K_{II}\,\gamma_{T_{HCO_3^-}}^2} \right)^{\frac{1}{2}} + \frac{a_{CO_3^{--}}}{\gamma_{T_{CO_3^{--}}}} \tag{13}$$

The above two equations are best solved for $a_{CO_3^{--}}$ by trial and error or through the use of an automatic computer, as is the case for the calculation of a_{CO_3} from A_C and ΣCO_2, using Eqs. (3), (4), (5), (1), and (2),

$$A_C - \Sigma CO_2 = \frac{a_{CO_3^{--}}}{\gamma_{T_{CO_3^{--}}}} - \left[\frac{K_{II}\left(A_C - \dfrac{2a_{CO_3^{--}}}{\gamma_{T_{CO_3^{--}}}} \right)^2 \gamma_{T_{HCO_3^-}}^2}{K_I\,a_{CO_3^{--}}\,\gamma_{H_2CO_3}} \right] \tag{14}$$

FRESH WATER

In fresh waters, where $\gamma_T \approx \gamma$ and the Debye-Hückel expression can be used, the state of $CaCO_3$ saturation is most commonly determined from A_C and pH.

Table 4-1   Calculation of IAP_{CaCO_3} for a stream water, Birch Creek, California. [After Barnes (1965)]

Chemical composition		IAP Calculation	
Ion	ppm		
		Temperature	20°C
Ca^{++}	73	Ionic strength	9.5×10^{-3}
Mg^{++}	21.0	$m_{HCO_3^-}$	4.0×10^{-3}
Na^+	14.0	$m_{Ca^{++}}$	1.8×10^{-3}
K^+	3.4	pH	8.60
HCO_3^-	245	$\gamma_{HCO_3^-}$	0.90
SO_4^{--}	93.0	$\gamma_{Ca^{++}}$	0.67
Cl^-	4.5	$a_{CO_3^{--}}$	6.6×10^{-5}
		$a_{Ca^{++}}$	1.2×10^{-3}
		IAP	7.9×10^{-8}
		$K_{aragonite}$	0.6×10^{-8}

Since most fresh waters have a pH less than 8.5, the value of A_C is equal to $m_{HCO_3^-}$. Thus, Eq. (10) can be used and the expression for IAP is

$$IAP = a_{Ca^{++}} a_{CO_3^{--}}$$
$$= \frac{\gamma_{Ca^{++}} \gamma_{CO_3} m_{Ca^{++}} A_C K_{II}}{a_{H^+}} \tag{15}$$

In some anaerobic fresh water the value of m_{HS^-} may approach that of $m_{HCO_3^-}$. In such case, the titration alkalinity includes HS^- and correction for it must be made to obtain $m_{HCO_3^-}$. No general statement can be made regarding the state of saturation of fresh river and lake water except that equilibrium ($IAP/K = 1$) is rare. Biological activity by aquatic organisms results in the production of CO_2 (respiration) or utilization of CO_2 (photosynthesis) so that P_{CO_2} of lake and river waters is usually different than P_{CO_2} in the surrounding air. Bacteria are especially important in raising P_{CO_2} and consequently lowering pH in deeper waters (see Chapter 7 for more details). As a result of biological controls on P_{CO_2} and pH, the value of IAP is perturbed from that expected for saturation with $CaCO_3$, and this explains the usual lack of equilibrium. An example for supersaturated river water is shown in Table 4-1.

SEA WATER

Sea water likewise is not in equilibrium with $CaCO_3$ due to the biological production and utilization of CO_2. The state of saturation of shallow ($P \approx 1$ atm), warm ($T \approx 25°C$) sea water of normal salinity can be determined directly from the Garrels and Thompson model discussed in Chapter 3.

From the data of Table 3-5:

$$IAP = 13.5 \times 10^{-9} \tag{16}$$

This is in good agreement with the calculation of IAP from pH and A using the equation

$$IAP = \frac{m_{T_{Ca}^{++}} A_C \gamma_{T_{Ca}^{++}}}{\dfrac{2}{\gamma_{T_{CO_3}^{--}}} + \dfrac{a_{H^+}}{\gamma_{T_{HCO_3}^-} K_{II}}} \tag{17}$$

and the values $m_{T_{Ca}^{++}} = 0.0103$, $A_C = 2.25 \times 10^{-3}$, pH $= 8.15$, $K_{II} = 10^{-10.33}$, $\gamma_{T_{Ca}^{++}} = 0.20$, $\gamma_{T_{CO_3}^{--}} = 0.021$, and $\gamma_{T_{HCO_3}^-} = 0.55$ (Berner, 1965):

$$IAP = 12.5 \times 10^{-9} \tag{18}$$

By comparison, from the tabulated data of ΔF_f° in Appendix I,

$$K_{calcite} = 4.0 \times 10^{-9} \tag{19}$$

$$K_{aragonite} = 6.3 \times 10^{-9} \tag{20}$$

Therefore, warm, shallow sea water is distinctly supersaturated. For calcite

$$IAP/K = 3.1 \tag{21}$$

Supersaturation by approximately this degree is also shown by the work of Weyl (1961). Weyl determined the change of pH upon addition of calcite to sea water. A distinct drop indicated precipitation on the added calcite and thus supersaturation.

Most of the oceans are not at 25°C and 1-atm total pressure. Thus, corrections for the effect of temperature and pressure upon $\gamma_{T_{HCO_3}^-}$, $\gamma_{T_{Ca}^{++}}$, $\gamma_{T_{CO_3}^{--}}$, K_{II}, and $K_{calcite}$ must be made in order to determine the state of saturation. In addition, pH varies with depth due to organic activity and *in situ* values must be known. Because of the relative constancy of salinity and composition of sea water, the assumption of constant $m_{T_{Ca}^{++}}$ and A_C throughout the oceans is good and the values similar to those above can be used. The other pertinent expressions from Chapter 2 are:

$$\left.\frac{d\log K}{dT}\right|_P = \frac{\Delta h^\circ}{2.3RT^2} \tag{22}$$

$$\left.\frac{d\log K}{dP}\right|_T = \frac{-\Delta v^\circ}{2.3RT} \tag{23}$$

$$\left.\frac{\partial\log \gamma_T}{\partial T}\right|_P = \frac{h^\circ - h}{2.3RT^2} \tag{24}$$

$$\left.\frac{\partial\log \gamma_T}{\partial P}\right|_T = \frac{v - v^\circ}{2.3RT} \tag{25}$$

where v and h refer to partial molal volume and enthalpy. The integrated expression for the effect of temperature on K_{calcite} at $P = 1$ atm is given by Berner (1965):

$$\log\left(\frac{K_T}{K_{298°}}\right)_{\text{calcite}} = 5860\left[\frac{1}{298} - \frac{1}{T}\right] + 50.2\log\frac{298}{T} \qquad (26)$$

This includes the assumption that $\Delta h°$ is a linear function of temperature. The integrated expression for the pressure effect on K_{calcite} is given by Owen and Brinkley (1941):

$$\log\left(\frac{K_P}{K_1}\right)_{\text{calcite}} = \frac{-\Delta v°}{189T}(P - 1) - 2.78 \times 10^{-7}$$
$$\times\left[3000(P - 1) - 2.07 \times 10^7\log\left(\frac{3000 + P}{3000}\right)\right] \qquad (27)$$

Values of $\Delta v°$ at different temperatures encountered in the ocean can be obtained from the data of Zen (1957). The effects of temperature and pressure on K_{II} result in similar expressions; from Harned and Scholes (1941)

$$\log\left(\frac{K_T}{K_{298°}}\right)_{\text{II}} = 4980\left[\frac{1}{298} - \frac{1}{T}\right] + 32.5\log\frac{298}{T} \qquad (28)$$

The expression for pressure from Owen and Brinkley (1941) is

$$\log\left(\frac{K_P}{K_1}\right)_{\text{II}} = \frac{-\Delta v°}{189T}[P - 1] - 1.54 \times 10^{-7}$$
$$\times\left[3000(P - 1) - 2.07 \times 10^7\log\left(\frac{3000 + P}{3000}\right)\right] \qquad (29)$$

The value of $\Delta v°$ for 25°C is given by Owen and Brinkley (1941) as -27.8 cm^3/mole. Values for other temperatures have not been accurately measured.

Measurements of $h - h°$ for sea water have not been made. Nevertheless, the effect of temperature on $\gamma_{T_{\text{Ca}^{++}}}$, $\gamma_{T_{\text{CO}_3}^{--}}$, and $\gamma_{T_{\text{HCO}_3}^-}$ has been calculated by Li (1967) using experimental measurements by Lyman (1956) and MacIntyre (1965) for carbonate equilibria in sea water at different temperatures. For the temperature range in the open ocean (0 to 30°C) there is negligible variation in $\gamma_{T_{\text{CO}_3}^{--}}$. The expressions for $\gamma_{T_{\text{HCO}_3}^-}$ and $\gamma_{T_{\text{Ca}^{++}}}$ using the data of Li (1967) and Berner (1965) are

$$\gamma_{T_{\text{HCO}_3}^-} = 0.55 - 0.003\,(298 - T) \qquad (30)$$

$$\gamma_{T_{\text{Ca}^{++}}} = 0.20 + 0.002\,(298 - T) \qquad (31)$$

The value of v for Ca^{++} has been measured by Duedall and Weyl (1967) in sea water of average salinity. From their data it can be shown that at all depths in the ocean the effect of pressure on $\gamma_{T_{\text{Ca}}^{++}}$ is negligible. The determination of $v_{HCO_3^-}$ by Duedall and Weyl is insufficiently accurate because of failure to separate completely $v_{HCO_3^-}$ from $v_{CO_3^{--}}$. Instead the pressure effect on $\gamma_{T_{HCO_3^-}}$ has been calculated from the Disteches' (1967) measurements of the pressure effect on carbonate equilibria in sea water. The resulting equation for 22°C is

$$\log\left(\frac{\gamma_{T_P}}{\gamma_{T_1}}\right)_{HCO_3^-} = 1.13 \times 10^{-4}(P-1) \tag{32}$$

The value of v for CO_3^{--} has not been measured in sea water because of experimental difficulties. However, the effect of pressure on $\gamma_{T_{CO_3^{--}}}$ at 22°C can also be calculated from the experimental work of Disteche and Disteche (1967). The resulting empirical equation is

$$\log\left(\frac{\gamma_{T_P}}{\gamma_{T_1}}\right)_{CO_3^{--}} = 3.76 \times 10^{-4}(P-1) \tag{33}$$

The value of IAP for any P and T in the oceans can be calculated using the above expressions, measured *in situ* pH, and the equation

$$IAP = \frac{2.47 \times 10^{-5}\gamma_{T_{Ca^{++}}}}{\dfrac{2}{\gamma_{T_{CO_3^{--}}}} + \dfrac{a_{H^+}}{\gamma_{T_{HCO_3^-}}K_{II}}} \tag{34}$$

This is Eq. (17) in which the values $m_{T_{Ca^{++}}} = 1.03 \times 10^{-2}$, $A_C = 2.4 \times 10^{-3}$ have been substituted.

To date, due to technical difficulties, only a few *in situ* pH measurements of deep sea water have been made, although the preliminary work of Ben-Yaakov and Kaplan (1968) points to future improvements in this field. Consequently *in situ* pH must be calculated from pH of deep sea water samples measured after bringing the samples up to the surface. An approximate correction equation based on laboratory pressure measurements (Culberson and Pytkowicz, 1968) is

$$pH = pH_S - 4.0 \times 10^{-4}P \tag{35}$$

where $pH_S = pH$ measured on a sample brought to the surface from a depth of corresponding pressure P.

The state of saturation with respect to calcite, $IAP/K_{calcite}$, for a sample of northern Pacific sea water is presented in Table 4-2 and Fig. 4-1. Included in Table 4-2 are the values of pH, K_{II}, $\gamma_{T_{CO_3^{--}}}$, $\gamma_{T_{HCO_3^-}}$, and $\gamma_{T_{Ca^{++}}}$ used in calculating IAP. The relation between depth and pressure is assumed to be 10 meters = 1 atm. It is apparent that below a few hundred meters northern Pacific sea water becomes undersaturated with respect to calcite. Since calcite is less soluble (more stable) than aragonite, the water is also undersaturated with respect to aragonite. The cause for the undersaturation is twofold. At shallower depths lowering of pH, due to biological CO_2

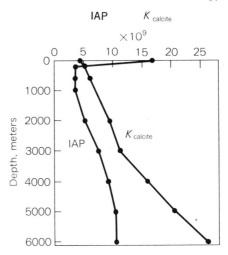

Fig. 4-1 Distribution of IAP and K_{calcite} with water depth for a station in the north Pacific Ocean. Note the greatly increasing degree of undersaturation below 4000 meters.

production, causes a considerable decrease in IAP. At greater depths increased pressure causes a large increase in K for calcite.

Undersaturation of the northern and east equatorial Pacific below a few hundred meters is also shown by the work of Pytkowicz (1965a), Berner (1965), Peterson (1966), Berger (1967), Lyakhin (1968), and Hawley and Pytkowicz (1969). The method used by Pytkowicz is similar to that used here except that actual measurements of $K_{m_{\text{calcite}}}$ in sea water as a function of temperature and pressure were used and compared to IMP. (The ratio IMP/K_m is equivalent to IAP/K.) Values of IMP were calculated from pH and A_C using measured values of the second dissociation constant of carbonic acid in sea water, $K_{\text{II}}' = m_{T_{\text{CO}_3}}{}^{--}a_{\text{H}^+}/m_{T_{\text{HCO}_3}}{}^{-}$ (Culberson and Pytkowicz, 1968). Berger and Peterson demonstrated undersaturation directly. They suspended samples of calcite (as polished spheres of iceland spar or as foraminiferal tests) at various depths on an anchored cable and measured the weight loss after exposure to subsurface sea water for several months. They found that the rate of solution increased rapidly below 3000 to 4000 meters but that solution also took place at shallower depths.

Li (1967, 1969), on the basis of equilibrium calculations, maintains that most of the ocean is not undersaturated with calcite. In the Atlantic he shows that sea water is undersaturated only below about 4000 meters. In the Pacific his data do not agree with that of Pytkowicz. He finds a lower degree of undersaturation or none where Pytkowicz finds distinct undersaturation. One possible cause of this discrepancy is that Li uses P_{CO_2} and ΣCO_2 to determine IMP, whereas Pytkowicz uses pH and A_C. Differences between the calculations in this chapter and those of Li can be explained on the basis of differences in γ_T values at 25°C and 1-atm pressure and the failure of Li to include the effect of temperature on $\partial \ln K/\partial P$. It can be shown that below the thermocline, Li's assumption that $\Delta v^\circ/T$ is the same as that for 25°C leads to

Table 4-2 State of saturation with calcite, IAP/$K_{calcite}$, for northern Pacific sea water

Temperatures are averages for latitudes 25 to 55° N. Values of pH are taken from Park (1966) and are corrected for pressure

Depth, meters	Temp., °C	pH	$\gamma_{TCO_3^-}$	$\gamma_{THCO_3^-}$	$10^{11} \times K_{II}$	$\gamma_{TCa^{++}}$	$10^9 \times IAP$	$10^9 \times K_{calcite}$	IAP/$K_{calcite}$
0	20	8.25	0.021	0.53	4.6	0.21	16.6	4.3	3.85
200	10	7.60	0.021	0.50	3.5	0.23	3.8	5.2	0.73
600	6	7.65	0.022	0.50	2.9	0.24	3.6	6.0	0.60
1000	3	7.65	0.023	0.49	2.8	0.245	3.6	7.0	0.51
2000	2	7.75	0.025	0.51	3.1	0.25	5.1	9.4	0.54
3000	2	7.85	0.027	0.52	3.5	0.25	7.4	12.1	0.61
4000	2	7.90	0.030	0.53	3.9	0.25	9.1	15.7	0.58
5000	2	7.90	0.0325	0.55	4.3	0.25	10.4	20.3	0.51
6000	2	7.85	0.035	0.56	4.7	0.25	10.5	25.9	0.41

a value of $K_{calcite}$ that is 20 percent too low at 4000 meters. Also, Li uses the values (at 25°C, $P = 1$ atm) $\gamma_{T_{Ca}^{++}} = 0.23$, $\gamma_{T_{CO_3}^{--}} = 0.031$, as compared to $\gamma_{T_{Ca}^{++}} = 0.20$, $\gamma_{T_{CO_3}^{--}} = 0.021$ used in the present study. The latter two values are in better agreement with independent studies (Garrels and Thompson, 1962; Thompson and Ross, 1966).

Because of the foregoing disagreements, nonstandardization of methods, and lack of extensive data, nothing definite can be said regarding the state of saturation of the oceans in general. However, all workers show distinct undersaturation of most of the water below the top few hundred meters in the north Pacific. In the south Pacific (excluding the eastern margin) and Atlantic Oceans supersaturation appears to extend to much greater depths as shown by recent work of Hawley and Pytkowicz (1969) and Li et al. (1969), but more work is needed to check their conclusions.

DISSOLUTION OF CaCO₃ IN THE OCEANS

In deep-sea pelagic sediments calcium carbonate is abundant, mainly in the form of calcitic foraminiferal tests, but only at more shallow depths (1000 to 4000 meters). Below ~4000 meters the content of calcium carbonate begins to decrease rapidly with depth, because of dissolution, so that very little is found below 5500 meters. This is shown in Fig. 4-2. The zone where $CaCO_3$ drops off rapidly is known as the *compensation depth* and represents the depth where $CaCO_3$ is dissolved as fast as it is deposited. It has been stated often that the compensation depth simply represents the equilibrium boundary where sea water passes downward from supersaturation to undersaturation. However, the whole concept of the compensation depth is a *kinetic* one con-

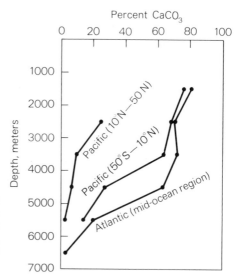

Fig. 4-2 Plot of percent CaCO₃ versus water depth for deep sea pelagic sediments. Values are averages for each 1000-meter depth interval. (*After Sverdrup, Johnson, and Fleming, 1942, revised to include the data of Turekian, 1964, for the Atlantic Ocean.*)

cerned with the rate of dissolution.   It is quite conceivable that fast deposition combined with slow dissolution may enable the accumulation of appreciable $CaCO_3$ in weakly undersaturated water.   This is suggested by the observation that the compensation depth is not a simple depth at all but often a rather wide depth zone.   In fact in the north Pacific a compensation depth cannot really be delineated (see Fig. 4-2), and $CaCO_3$ persists at depths far below that where the water becomes distinctly undersaturated.

There is no doubt that the rate of dissolution of $CaCO_3$ increases with depth in the ocean due to increasing undersaturation.   This has been demonstrated by Peterson (1966) and Berger (1967), who suspended calcite spheres and foram samples respectively in Pacific sea water at various depths and

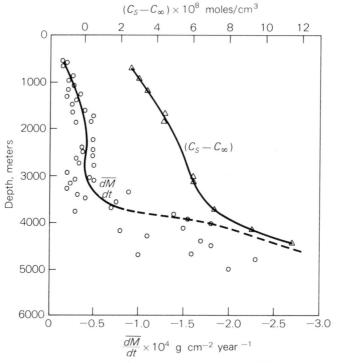

Fig. 4-3 Plot of rate of dissolution of calcite spheres dM/dt (see text) and degree of undersaturation ($C_S - C_\infty$) versus water depth for Pacific Ocean water at about 20°N, 170°W.   The curve for dM/dt is drawn through averages for five adjacent measurement points (*Peterson, 1966*). Five high rate of dissolution points at 4000- to 5000-meter depth fall to the right of the diagram and have been omitted.   Values of ($C_S - C_\infty$) correspond to the increase in concentration of $m_{T_{CO_3^{--}}}$ if each water sample were brought to equilibrium with calcite.   The values are based on state of saturation calculations, using the methods of the present chapter.

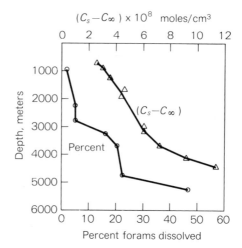

Fig. 4-4 Plot of percent foraminifera dissolved in four months (*Berger, 1967*) and degree of undersaturation $C_S - C_\infty$ versus water depth for Pacific Ocean water at about 20°N, 170°W. Values of $C_S - C_\infty$ are the same as in Fig. 4-3.

measured weight loss with time. Their results are shown in Figs. 4-3 and 4-4. Added to the figures is the degree of undersaturation of sea water for the same general area where measurements were obtained by Peterson and by Berger. The general parallelism of the curves suggests a direct relation between the rate of dissolution and degree of undersaturation. This would be expected if dissolution were controlled by diffusion.

A test of the hypothesis that dissolution is diffusion-controlled is provided by Peterson's experiment, where calcite spheres were used. The rate of dissolution of spheres by diffusion is given by Eq. (88) of Chapter 2 as

$$\frac{-dr}{dt} = \frac{vD(C_S - C_\infty)}{r} \tag{36}$$

In terms of mass, per unit area, $\bar{M}$:

$$\frac{\overline{dM}}{dt} = \frac{-D(C_S - C_\infty)}{r} \tag{37}$$

The parameter $(C_S - C_\infty)$ is assumed to be the number of moles of dissolved calcium carbonate needed to bring the water to saturation with calcite. This can be calculated from both Soviet and American pH and alkalinity data for subsurface water taken within a few degrees of latitude and longitude of the site of Peterson's experiment. In addition, data of Li (1967) for P_{CO_2} and ΣCO_2 in the waters of the same general region are also available.

Average values of radius r and $\overline{dM}/dt$ measured by Peterson (see Fig. 4-3) when combined with $(C_S - C_\infty)$ enable calculation of the diffusion coefficient D for each corresponding water depth via Eq. (37). Resulting values range from 0.015 to 0.06 $\times$ 10^{-5} cm²/sec if the methods of the present chapter for determining the state of saturation of sea water are used. Use of Li's data and Li's or

Pytkowicz's method of calculation enables calculation of the *maximum* value for D at any depth of 0.18×10^{-5} cm²/sec. For any given calculation method values of D increase severalfold with depth. Variability of D with depth and the values shown above indicate that dissolution of Peterson's spheres was not controlled by simple ionic diffusion. The absolute minimum value of D in sea water should be about 1.0×10^{-5} cm²/sec, which corresponds to simple diffusion without turbulent mixing. The values of D calculated from Eq. (37) are an order of magnitude less than this value. This suggests that dissolution is inhibited by some process taking place at the calcite surface. The process may be protective armoring by polar organic molecules (Smith et al., 1968) or selective adsorption of Mg^{++} (Weyl, 1965; Berner, 1967). Both are known to decelerate the dissolution of calcite.

Once a calcareous test reaches the sea bottom it may be further protected from dissolution by a partly saturated boundary layer at the sediment-water interface. In this case the rate of dissolution would be limited by the very slow rate at which dissolved calcium carbonate can diffuse out of the sediment as a whole. Thus, the measured rates obtained by Peterson and Berger are probably far too high for calcareous remains resting on the bottom.

PRECIPITATION OF CaCO₃ IN THE OCEANS

Data presented earlier in this chapter indicate that most surface sea water is supersaturated with respect to both calcite and aragonite. Therefore, one might expect to find widespread $CaCO_3$ precipitation in areas of shallow water, such as the continental shelves. The geological evidence does not bear this out. Precipitation believed to be inorganic is mostly confined to subtropical and tropical environments, where it is overall quantitatively unimportant, compared to biological secretion, as a source of supply of $CaCO_3$ to sediments.

The best known and most widely accepted example of an inorganic precipitate formed from overlying sea water is aragonite oolites (Newell et al., 1960). Oolites consist of crystals of aragonite surrounding a central nucleus, the resulting form being a sand-sized spherical body containing concentric shells of aragonite (see Fig. 4-5). Oolites occur only in or near areas of strong wave and current action, where they may accumulate to form beach sands and channel or barrier bars. The formation of oolites is dependent upon suspension combined with heterogeneous nucleation. A biogenic calcium carbonate fragment usually acts as the initial nucleus. Once a layer of aragonite is formed on the nucleus, further precipitation is aided by the aragonite crystals which may act as seeds. Precipitation does not occur, however, while the oolites are resting on the bottom. The pore water, because of its semiclosed nature (see Chapter 1), rapidly equilibrates with aragonite and no further precipitation takes place. This is where suspension becomes important. The overlying water is always supersaturated so that when oolites are thrown up

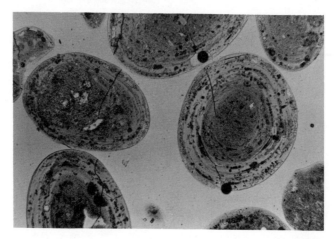

Fig. 4-5 Photomicrograph of mounted and sectioned modern aragonite oolites from the Bahama Banks. Transmitted light (× 200). Note the concentric banding due to tangential orientation of aragonite needles (the two large black dots are bubbles in the mounting).

into it due to turbulence, further aragonite crystallization on the oolites can occur. Because growth takes place during suspension, a spherical form results; i.e., the water is equally supersaturated on all sides of the oolite. The inclusion of organic matter within specific aragonite layers probably represents algal growths on the oolites during periods of quiescence (Newell et al., 1960). Organic matter is not necessary for oolite formation, as suggested by some writers. This is shown by the formation of artificial oolites from hard water in boilers (Twenhofel, 1928), a process physically analogous to natural oolite growth.

Another example of inorganic precipitation from sea water is the formation of beachrock and subtidal carbonate cements. The precipitation is heterogeneous on pre-existing detritus, and presumably results from the flow of supersaturated sea water through the sediments. Since cementation is basically a subsurface process, further discussion of this topic is postponed to later chapters on diagenesis.

Many supposed instances of sudden massive inorganic calcium carbonate precipitation from sea water have been cited, the classic example being the quiet shallow waters west of Andros Island on the Bahama Banks (Smith, 1940; Cloud, 1962). In this region colder sea water from the surrounding oceanic deeps enters the shallow bank region and is warmed and undergoes evaporation as it slowly flows across the banks. As a result, there is an increase in salinity inward toward the west shore of Andros Island. Accompanying the increase in salinity and warming there is a loss of $CaCO_3$ from the water as indicated by

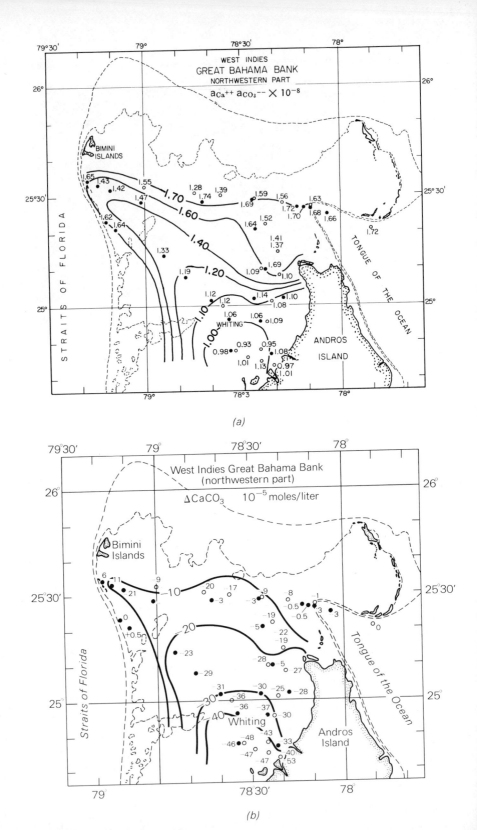

(a)

(b)

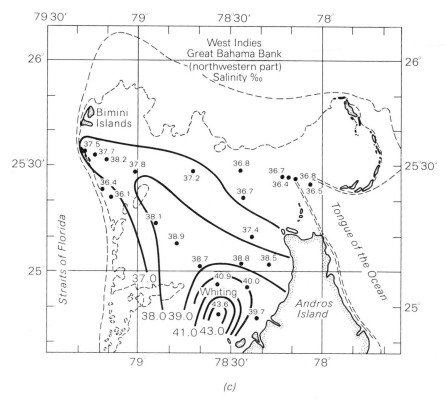

(c)

Fig. 4-6 Maps of the Northwestern part of the Bahama Banks (*after Broecker and Takahashi, 1966*) showing (*a*) ion-activity product contours for calcium carbonate; (*b*) contours of loss of $CaCO_3$ from solution, $\Delta CaCO_3$; (*c*) salinity contours.

a drop in carbonate alkalinity and in IAP.   This is all shown in Fig. 4-6.   The local rate of loss of $CaCO_3$ has been calculated by Broecker and Takahashi (1966) by dividing the $CaCO_3$ deficit by the *mean residence time* on the Banks for each water sample.   Mean residence time was calculated by the degree of uptake of excess bomb-produced C^{14} from the atmosphere.   As a result of volume-averaging, a mean value of precipitation rate of 50 mg $CaCO_3$ cm^{-2} $year^{-1}$ was obtained for the area shown in Fig. 4-6.   This compares favorably with values calculated from the rate of accumulation of $CaCO_3$ in the underlying sediments (Cloud, 1962).

The characteristic sediment found on the banks west of Andros Island is a fine-grained mud made up of micron-sized needle-shaped crystals of aragonite.   Aragonite of the same size and morphology can be precipitated from sea water in the laboratory, and this has led to the speculation that the aragonite is inorganic.   Examples of supposed massive inorganic precipitation actually in progress have been cited in the form of *whitings*.   Whitings are localized patches of water, whitened by the presence of suspended aragonite, which occasionally appear on the banks. Broecker and Takahashi (1966), however, have conclusively demonstrated that at least some whitings result from the resuspension of bottom sediment and not from precipitation. They showed that the C^{14}/C^{12} ratio in the suspended aragonite of a whiting was similar to that in the underlying sediment and completely different from the ratio in the water of the whiting. Also, the whiting water did not show a loss of dissolved $CaCO_3$ anywhere near that needed to account for the suspended aragonite. This conclusion is in agreement with that of field workers who have observed whitings produced by the stirring activity of schools of bottom-feeding fish.

In opposition to the idea of inorganic precipitation is the theory that the aragonite needle muds on the Bahama Banks result from the disaggregation of calcareous algae. Algae which secrete aragonite needles very similar to those found in the sediments commonly occur on the Banks. Lowenstam and Epstein (1957) have shown that the algae and sediment needles are very similar in their isotopic ratios C^{13}/C^{12} and O^{18}/O^{16}. From this they have concluded that the needle muds are biogenic in origin. Cloud (1962), however, has pointed out that inorganic aragonite precipitated during summer months from the same sea water where the algae are found could also show a similar isotopic composition so that isotopic ratios are not convincingly diagnostic of algal secretion. A better argument is whether the rate of algal secretion is quantitatively sufficient to account for the measured rate of aragonite precipitation. Recent estimates of Neumann and Land (1968), based on studies of a nearby area, indicate that algae can produce more than enough aragonite to qualify as a source for all the needles in the sediments. Thus, the evidence for an inorganic origin of the needles is equivocal.

To the writer's knowledge the unseeded, nonbiological precipitation of aragonite in the laboratory from untreated normal pH sea water has not been satisfactorily demonstrated. Reported precipitations have been achieved either by the addition of alkali, which increases the carbonate alkalinity, or by the removal of $CO_{2\,aq}$, which raises the pH (Revelle and Fleming, 1934; Cloud, 1962, p. 76). (In Cloud's experiment, which was supposed to be CO_2 removal only, the carbonate alkalinity could have inadvertently been increased by the slow transport of NaOH during the bubbling through sea water of NaOH-treated, CO_2-free air.) Precipitation by the addition of sodium carbonate has been studied by Pytkowicz (1965b). He found an exponential increase in the induction period as the amount of added carbonate was decreased. Extrapolation of his results to zero-added carbonate gave an induction period of many thousands of years. On this basis Pytkowicz explained why inorganic precipitation from average surface sea water would be too slow to be of any importance. Unfortunately Pytkowicz's experiments may pertain only to his experimental apparatus. This is because dust particles or the coated or uncoated surfaces of laboratory glassware probably induce heterogeneous nucleation and these nucleating agents are not representative of suspended nucleating particles in natural waters. The present author has personally found that precipitations of this type are highly irreproducible.

It is most unlikely that *homogeneous* nucleation of $CaCO_3$ occurs either in natural sea water or in experiments like those of Pytkowicz. This can be shown through the use of Eq. (87) of Chapter 2:

$$N = \bar{\nu} \exp \left[\frac{-4b^3 \sigma^3 v_a^2}{27 \oplus^2 \mathscr{K} T} \right] \tag{38}$$

In order to obtain measurable rates of nucleation N for $CaCO_3$ with $b = 6$, $v_a = 6 \times 10^{-23}$ cm^3, and $\sigma \approx 250$ ergs/cm^2 (Chave and Schmalz, 1966), at room temperature the degree of supersaturation IAP/K must be on the order of one billion. Since nucleation occurs at much lower degrees of supersaturation, it must be heterogeneous.

One important experimental observation made by Pytkowicz, which can be directly applied to the oceans, is that the rate of nucleation of $CaCO_3$ in sea water is greatly retarded by the presence of Mg_{aq}^{++}. The probable reason for this is that Mg^{++} and Ca^{++} compete for crystallographic sites in growing carbonate crystal embryos and as a result of Mg uptake many embryos are destabilized and redissolved before the critical nucleus size is reached. (The role of Mg^{++} in $CaCO_3$ crystallization is further discussed in Chapter 8.) It is likely that the persistence of $CaCO_3$ supersaturation in shallow sea water and the consequent lack of evidence for widespread precipitation are due at least partly to the inhibiting effect of dissolved Mg^{++} on nucleation rates. A larger degree of supersaturation is necessary for massive precipitation to occur at the concentration of Mg^{++} in average sea water.

Another factor possibly inhibiting precipitation is the presence in sea water of dissolved and suspended organic matter. Suess (1970) has shown that small traces of dissolved polar organic molecules in sea water are sufficient to bring about extensive adsorption on suspended $CaCO_3$. As a result an effective blocking of nucleation sites occurs and seeded precipitation is retarded. In the absence of $CaCO_3$ seeds it is also likely that dissolved organic matter interferes with nucleation and growth. The author has encountered silicate-mud pore waters very rich in dissolved organic matter in which the calcium carbonate equilibrium-ion product is greatly exceeded ($IAP/K \approx 15$). Yet precipitation of $CaCO_3$ does not take place even when the pore water is separated from the sediment and the supersaturation further increased.

REFERENCES

Barnes, Ivan, 1965, Geochemistry of Birch Creek, Inyo County, California, a travertine depositing creek in an arid climate, *Geochim. et Cosmochim. Acta*, v. 29, pp. 85–112.

Ben-Yaakov, S., and Kaplan, I. R., 1968, High pressure pH sensor for oceanographic applications, *Rev. Sci. Instruments*, v. 39, pp. 1133–1138.

Berger, W. H., 1967, Foraminiferal ooze: solution at depth, *Science*, v. 156, pp. 383–385.

Berner, R. A., 1965, Activity coefficients of bicarbonate, carbonate, and calcium ions in sea water, *Geochim. et Cosmochim. Acta*, v. 29, pp. 947–965.

————, 1967, Comparative dissolution characteristics of carbonate minerals in the presence and absence of aqueous magnesium ion, *Am. Jour. Sci.*, v. 265, pp. 45–70.

Broecker, W. S., and Takahashi, T., 1966, Calcium carbonate precipitation on the Bahama Banks, *Jour. Geophys. Research*, v. 71, pp. 1575–1602.

Chave, K. E., and Schmalz, R. F., 1966, Carbonate–sea water interactions, *Geochim. et Cosmochim. Acta*, v. 30, pp. 1037–1048.

Cloud, P. E., 1962, Environment of calcium carbonate deposition west of Andros Island, Bahamas, *U.S. Geol. Survey Prof. Paper* 350, 138 p.

Culberson, C., and Pytkowicz, R. M., 1968, Effect of pressure on carbonic acid, boric acid, and the pH in sea water, *Limn. and Oceanog.*, v. 13, pp. 403–417.

Disteche, A., and Disteche, S., 1967, The effect of pressure on the dissociation of carbonic acid from measurements with buffered glass electrode cells, *Jour. Electrochem. Soc.*, v. 114, pp. 330–340.

Duedall, I. W., and Weyl, P. K., 1967, The partial equivalent volumes of salts in seawater, *Limn. and Oceanog.*, v. 12, pp. 52–59.

Garrels, R. M., and Thompson, M. E., 1962, A chemical model for sea water at 25°C and one atmosphere total pressure, *Am. Jour. Sci.*, v. 260, pp. 57–66.

Harned, H. S., and Scholes, S. R., 1941, The ionization constant of HCO_3^- from 0 to 50°C., *Jour. Am. Chem. Soc.*, v. 63, pp. 1706–1709.

Hawley, J., and Pytkowicz, R. M., 1969, Solubility of calcium carbonate in seawater at high pressures and 2°C, *Geochim. et Cosmochim. Acta*, v. 33, pp. 1557–1560.

Li, Y-H., 1967, "The degree of saturation of $CaCO_3$ in the oceans," Ph.D. thesis, Columbia University, New York, 176 p.

————, Takahashi, T., and Broecker, W. S., 1969, Degree of saturation of $CaCO_3$ in the oceans, *Jour. Geophys. Research*, v. 74, pp. 5507–5525.

Lowenstam, H. A., and Epstein, S., 1957, On the origin of sedimentary aragonite needles of the Great Bahama Bank, *Jour. Geol.*, v. 65, pp. 364–375.

Lyakhin, Y. I., 1968, Calcium carbonate saturation of Pacific Water, *Oceanology*, v. 8, pp. 44–53.

Lyman, J., 1956, Buffer mechanism of sea water, Ph.D. thesis, University of California, Los Angeles, 198 p.

MacIntyre, W. G., 1965, The temperature variation of the solubility product of calcium carbonate in sea water, *Fisheries Research Board Canada Rept. Series No.* 200, 153 p.

Neumann, A. C., and Land, L. S., 1968, Algal production and lime mud deposition in the Bight of Abaco: A Budget, *Geol. Soc. Am. Abstracts of Annual Meetings*, Mexico City, p. 219.

Newell, N. D., Purdy, E. G., and Imbrie, J., 1960, Bahamian oölitic sand, *Jour. Geol.*, v. 68, pp. 481–497.

Owen, B. B., and Brinkley, S. R., 1941, Calculation of the effect of pressure upon ionic equilibria in pure water and in salt solutions, *Chem. Rev.*, v. 29, pp. 461–474.

Park, K., 1966, Deep sea pH, *Science*, v. 154, pp. 1540–1541.

Peterson, M. N. A., 1966, Calcite: rates of dissolution in a vertical profile in the central Pacific, *Science*, v. 154, pp. 1542–1544.

Pytkowicz, R. M., 1965a, Calcium carbonate saturation in the ocean, *Limn. and Oceanog.*, v. 10, pp. 220–225.

————, 1965b, Rates of inorganic calcium carbonate nucleation, *Jour. Geol.*, v. 73, pp. 196–199.

Revelle, R., and Fleming, R. H., 1934, The solubility product constant of calcium carbonate in sea water, *Pacific Sci. Cong. Proc.*, 5th, *Victoria and Vancouver, Canada*, v. 3, pp. 2089–2092.

Smith, C. L., 1940, The Great Bahama Bank; II. Calcium carbonate precipitation, *Jour. Marine Res.*, v. 3, pp. 171–189.

Smith, S. V., Dygas, J. A., and Chave, K. E., 1968, Distribution of calcium carbonate in pelagic sediments, *Marine Geology*, v. 6, pp. 391–400.

Suess, E., 1970, Interaction of organic compounds with calcium carbonate; I. Association phenomena and geochemical implications, *Geochim. et Cosmochim. Acta*, v. 34, pp. 157–168.

Sverdrup, H. U., Johnson, M. W., and Fleming, R. H., 1942, *The oceans; their physics, chemistry, and general biology*, Prentice-Hall, Englewood Cliffs, N.J., 1087 p.

Thompson, M. E., and Ross, J. W., 1966, Calcium in sea water by electrode measurement, *Science*, v. 154, pp. 1643–1644.

Turekian, K. K., 1964, The geochemistry of the Atlantic Ocean basin, *Trans. New York Acad. Sci.*, v. 26, pp. 312–330.

Twenhofel, W. H., 1928, Oölites of artificial origin, *Jour. Geol.*, v. 36, pp. 564–568.

Weyl, P. K., 1961, The carbonate saturometer, *Jour. Geol.*, v. 69, pp. 32–44.

————, 1965, The solution behavior of carbonate materials in sea water, *Proceedings of the International Conf. on Tropical Oceanog.*, Miami Beach, Florida, pp. 178–228.

Zen, E–an, 1957, Partial molar volumes of some salts in aqueous solutions, *Geochim. et Cosmochim. Acta*, v. 12, pp. 103–122.

5
Evaporite Formation

By far the most abundant evaporite minerals found in sedimentary rocks are halite, anhydrite, and gypsum. This chapter is concerned with the physics and chemistry of formation of these three minerals from sea water.

THE EVAPORATION OF SEA WATER

Sea water of normal salinity is undersaturated with respect to gypsum, anhydrite, and halite. This can be seen by comparison of IAP with K for each mineral. From Chapter 3, Table 3-5, the IAP values for sea water of $I = 0.7$ are

$$IAP_{CaSO_4} = a_{Ca^{++}} a_{SO_4^{--}} = 4.6 \times 10^{-6} \tag{1}$$

$$IAP_{NaCl} = a_{Na^+} a_{Cl^-} = 0.12 \tag{2}$$

By comparison the corresponding values of K for anhydrite ($CaSO_4$) and halite ($NaCl$) at 25°C, calculated from the $\Delta F°$ data of Appendix I, are

$$K_{anhydrite} = 4.2 \times 10^{-5} \tag{3}$$

$$K_{halite} = 38 \tag{4}$$

For gypsum, $CaSO_4 \cdot 2H_2O$, the expression for IAP is

$$IAP_{gypsum} = a_{Ca^{++}} a_{SO_4^{--}} a_{H_2O}^2 \qquad (5)$$

In sea water $a_{H2O} = 0.98$ and for the present purposes can be assumed to be equal to one. Therefore

$$IAP_{gypsum} = 4.6 \times 10^{-6} \qquad (6)$$

The value of K calculated from thermodynamic data is

$$K_{gypsum} = 2.5 \times 10^{-5} \qquad (7)$$

Notice that halite is more undersaturated than gypsum or anhydrite. Upon evaporation of sea water and consequent increase in IAP, gypsum (or anhydrite) would thus be expected to begin to precipitate earlier than halite. This is found to be the case if sea water is simply allowed to evaporate at room temperature; gypsum precipitation is followed at a more advanced stage of evaporation by halite precipitation.

The sequence of precipitation of salts from sea water during evaporation was originally worked out by Usiglio (1849). He found that gypsum first appeared after normal sea water had been evaporated to 19 percent of its original volume. The corresponding density change was from 1.025 to 1.126. Halite first appeared when the water had reached 9.5 percent of its original volume and the density was 1.214. Anhydrite precipitation was not observed. Although evaporation was relatively slow these values do not necessarily represent the stages of evaporation where IAP = K since some supersaturation is necessary to enable nucleation and precipitation. For gypsum the stage of evaporation where IAP = K was determined by Posnjak (1940). Posnjak added gypsum to calcium-free artificial sea water which had been evaporated to varying degrees and recorded the $CaSO_4$ concentration at saturation. The degree of evaporation at which the saturation concentration of $CaSO_4$ is the same as that of normal sea water evaporated to the same extent should be the point where IAP = K. This was found to occur at 30°C, when sea water was evaporated to 30 percent of its original volume. Thus, the value of 19 percent of original volume observed by Usiglio for the first appearance of gypsum is not the equilibrium value.

The degree of evaporation necessary to reach halite saturation can be calculated by means of Harned's rule and the equilibrium relation:

$$K_{halite} = \gamma_{\pm NaCl}^2 \, m_{Na_T^+} m_{Cl^-} \qquad (8)$$

If sea water is considered as an electrolyte mixture of NaCl, Na_2SO_4, $MgCl_2$, $CaCl_2$, and KCl, Harned's rule from Chapter 2 is

$$\log \gamma_{\pm NaCl} = \log \gamma_{\pm NaCl(0)} + \alpha_A I_A + \alpha_B I_B + \alpha_C I_C + \alpha_D I_D \qquad (9)$$

where $A = Na_2SO_4$, $B = MgCl_2$, $C = CaCl_2$, and $D = KCl$. In sea water up to the point of saturation with halite

$$\frac{m_{T_{Na^+}}}{m_{Cl^-}} = 0.86 \tag{10}$$

Therefore, substituting Eqs. (9) and (10) in Eq. (8) for evaporated sea water in equilibrium with halite

$$\log m_{Cl^-} = 0.033 + \tfrac{1}{2}\log K_{halite} - \log \gamma_{\pm NaCl(0)} - \alpha_A I_A - \alpha_B I_B$$
$$- \alpha_C I_C - \alpha_D I_D \tag{11}$$

In order to obtain the various I values and $\gamma_{\pm NaCl(0)}$ in Eq. (11), as a first approximation evaporation is assumed to be to 10.0 percent the original sea water volume. At this degree of evaporation the total ionic strength, after correcting for $CaSO_4$ precipitation, is 7.5. The value of $\gamma_{\pm NaCl(0)}$ can be calculated for $I_T = 7.5$ by extrapolation from measured values at lower ionic strengths using a slightly corrected version of the Åkerlöf-Thomas rule (Robinson and Stokes, 1959, p. 447). Substitution in Eq. (11) of values for α from Fig. 3-1 and for I_A, I_B, I_C, I_D, $\gamma_{\pm NaCl(o)}$, and K (obtained from the free-energy data of Appendix I) enables calculation of a value for m_{Cl^-}. This value is then used to recompute the degree of evaporation and the above procedure is repeated until a constant value of m_{Cl^-} is obtained. The result is

$$m_{Cl^-} = 6.06 \tag{12}$$

which is the molality of chloride ion at the point of saturation. The value of m_{Cl^-} in normal sea water of salinity 35 per mill is 0.555. Thus, to attain a molality of 6.06 at a density of 1.214, sea water needs to be evaporated to 10.2 percent of its original volume. This is relatively close to the value 9.5 percent found by Usiglio for the onset of halite precipitation. The apparently low degree of supersaturation required to bring about halite precipitation suggests a dislocation-controlled growth mechanism (see Chapter 2).

GYPSUM AND ANHYDRITE

EQUILIBRIUM RELATIONS

The equilibrium constant at 1-atm total pressure for the reaction

$$CaSO_4 \cdot 2H_2O_{gypsum} \leftrightharpoons CaSO_{4\,anhydrite} + 2H_2O$$

has been measured accurately over a range of temperatures by Hardie (1967). Since neither gypsum nor anhydrite shows appreciable solid solution, K for this reaction is simply $a_{H_2O}^2$. Figure 5-1, a plot of a_{H_2O} for gypsum-anhydrite equilibrium versus temperature, is based on Hardie's data. The activity of

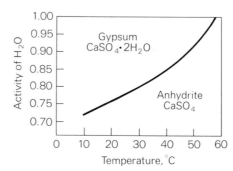

Fig. 5-1 Stability fields of gypsum and anhydrite at 1-atm total pressure. (*After Hardie, 1967.*)

water in sea water at the point where gypsum becomes saturated is approximately 0.93.   From Fig. 5-1 it can be seen that at this water activity gypsum is more stable than anhydrite at all temperatures below 50°C.   Therefore, at room temperature gypsum should be the first calcium sulfate mineral to precipitate from evaporating sea water and this is found to be the case as discussed earlier.   At 25°C anhydrite becomes stable when the activity of water is lowered to 0.78 but it has never been observed to precipitate in the laboratory by evaporation at any activity of water.   Thus, in highly saline waters gypsum is metastable.   At the point where sea water reaches halite saturation, the value of $a_{H_2O} = 0.74$.   From Fig. 5-1, it can be seen that at this water activity gypsum is stable only below 18°C.   For temperatures greater than 18°C the assemblage halite plus gypsum is metastable (at a pressure of one atmosphere).

The effect of pressure due to burial on the gypsum-anhydrite equilibrium has been calculated by MacDonald (1953) for two situations: one in which the pressure on all phases is lithostatic, and the other in which pressure on the minerals is lithostatic and that on the pore waters is hydrostatic.   The relation between lithostatic and hydrostatic pressure for a rock density of 2.4 is

$$P_{\text{lithostatic}} = 2.4 P_{\text{hydrostatic}} \tag{13}$$

Now from Appendix III, Eq. (8), denoting $P_{\text{lithostatic}}$ simply by P,

$$dF_{\text{gypsum}} = v_{\text{gypsum}} \, dP - s_{\text{gypsum}} \, dT \tag{14}$$

$$dF_{\text{anhydrite}} = v_{\text{anhydrite}} \, dP - s_{\text{anhydrite}} \, dT \tag{15}$$

For the lithostatic only situation

$$dF_{H_2O} = v_{H_2O} \, dP - s_{H_2O} \, dT \tag{16}$$

and for hydrostatic plus lithostatic situation

$$dF_{H_2O} = \frac{v_{H_2O}}{2.4} \, dP - s_{H_2O} \, dT \tag{17}$$

For the gypsum → anhydrite + 2 water reaction, the lithostatic situation yields

$$d(\varDelta F) = \varDelta v\, dP - \varDelta s\, dT \qquad\qquad (18)$$

and hydrostatic plus lithostatic situation yields

$$d(\varDelta F) = \left(v_{\text{anhydrite}} - v_{\text{gypsum}} + \frac{v_{\text{H}_2\text{O}}}{1.2}\right) dP - \varDelta s\, dT \qquad (19)$$

At equilibrium $d(\varDelta F) = 0$. Thus, for equilibrium:

$$\frac{dP}{dT} = \frac{\varDelta s}{\varDelta v} \qquad \text{(lithostatic)} \qquad\qquad (20)$$

$$\frac{dP}{dT} = \frac{\varDelta s}{[v_{\text{anhydrite}} - v_{\text{gypsum}} + (v_{\text{H}_2\text{O}}/1.2)]} \qquad \text{(hydrostatic and lithostatic)} \qquad (21)$$

MacDonald calculated P-T equilibrium curves from s and v data for the three phases for two situations, one for pure water ($a_{\text{H}_2\text{O}} = 1$) and the other for

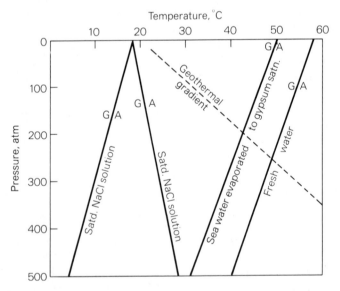

Fig. 5-2 Stability fields of gypsum (G) and anhydrite (A) as a function of temperature and pressure. The G–A curves with negative slope correspond to cases where pressure on the solids is lithostatic and that on the surrounding water is hydrostatic. The G–A curve with positive slope is for the pure lithostatic situation. The geothermal gradient is that measured over the Grand Saline Dome, Texas. (*After MacDonald, 1953, and Hardie, 1967.*)

halite saturation at the given P and T.   Modifications of his results to include new values of Δs and surface temperature based on the data of Hardie (1967) have been made in constructing Fig. 5-2.   Also shown is a geothermal gradient for an evaporite sedimentary terrain (Grand Saline Dome, Texas) taken from MacDonald's paper and a P-T equilibrium curve drawn for the activity of water at which gypsum saturation is reached in sea water.

ANHYDRITE FORMATION

The most important parameters controlling the relative stability of anhydrite and gypsum are temperature and salinity (a_{H_2O}).   As can be seen in Fig. 5-2, if anhydrite is not stable at the time of deposition, it becomes stable upon burial. The hydrostatic-lithostatic P-T equilibrium curve for sea water evaporated to $a_{H_2O} = 0.925$ (the minimum amount of evaporation necessary to saturate sea water with a $CaSO_4$ mineral at 30°C) intersects the geothermal gradient curve at $P = 200$ atm, corresponding to a depth of 833 meters.   For more concentrated brines the intersection depth is shallower and at saturation with NaCl anhydrite is stable right up to the surface.   Thus, gypsum, if precipitated originally from sea water, has a small stability field in the subsurface.   Most ancient evaporites, which have not been uplifted into zones of meteoric waters, contain anhydrite rather than gypsum.   This is in agreement with thermodynamic predictions.

During precipitation, gypsum rather than anhydrite forms, even when anhydrite is far more stable under the existing conditions.   The metastable coexistence of halite and gypsum in modern sediments is relatively common (Phleger, 1969).   Also, Hardie (1967) has demonstrated that gypsum can be readily converted in the laboratory to anhydrite; but that the direct precipitation of anhydrite cannot be achieved, even in the presence of anhydrite seeds. Therefore, it appears that most anhydrite forms originally as gypsum.   This helps to explain the rarity of "primary" anhydrite in modern sediments. One of the very few examples is that which occurs with dolomite in the mud flats (sabkhas) of the Trucial Coast of Arabia (Kinsman, 1966; 1969). Kinsman distinguishes two types of anhydrite: that which replaces gypsum as pseudomorphic crystals and that which constitutes interstitial fillings between pseudomorphs.   The latter occurrence he regards as a primary precipitate. However, in light of Hardie's experiments it is possible that both the pseudomorphs and interstitial fillings are the result of the dehydration of gypsum to anhydrite and that the sabkha anhydrite is an early diagenetic replacement of gypsum.

HALITE

The most abundant evaporite mineral in sedimentary rocks is halite, NaCl. As shown earlier, halite will crystallize from sea water after evaporation has proceeded to about 9 to 10 percent of the original water volume. Little metastable supersaturation is encountered, no metastable solids form as precursors, and large crystals form easily. Thus, halite is thermodynamically "well behaved," which is in distinct contrast to almost all other common sedimentary minerals.

The characteristic form adopted by halite during crystallization from an evaporating brine is the pyramidal hopper crystal. Dellwig (1955) has described the formation of salt hopper crystals as follows: during evaporation a very narrow supersaturation film forms at the air-brine interface within which halite is nucleated. When each crystal grows large enough, it begins to sink downward out of the supersaturation film so that growth can continue only upward and outward (see Fig. 5-3). Continued sinking and upward and outward growth result in a hollow inverted pyramid- or hopper-shaped crystal. Eventually the hopper crystal is disturbed by wave action, etc., so that it falls completely out of the supersaturation film and settles to the bottom of the brine body.

While on the bottom and after burial additional halite precipitation or recrystallization of the hopper crystals can take place. The result is over-growths, fillings, and replacements of the cloudy hopper crystals by clear crystals of secondary halite. Thus, hopper crystals may be used as a crude measure of the proportion of primary evaporitic halite present in a salt bed. Salt beds which have undergone extensive diagenetic recrystallization should be relatively free of hopper crystals.

In shallow water rapid changes in salinity or temperature should cause more extensive recrystallization of freshly deposited hopper crystals than in

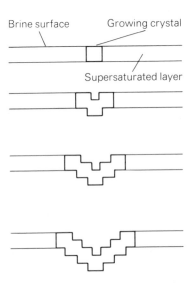

Fig. 5-3 Growth of a pyramidal hopper crystal of halite by evaporation. As the initial crystal (top) begins to sink out of the super-saturated layer, further growth is upward and outward only. The result is a hollow inverted pyramid or hopper shown in section at the bottom. (*After Dellwig, 1955.*)

deeper water where bottom conditions are more constant.   If so, it might be possible to use the abundance of hopper crystals in an ancient halite bed as a crude indicator of relative water depth.   The observations by Dellwig (1955) of the Silurian salt deposits of the Michigan Basin tend to confirm this hypothesis, although later work by the same author (Dellwig and Evans, 1969) is less conclusive.   More work on this subject is needed.

DEPOSITIONAL MODELS

Based on geological and oceanographic observations some basic physical models of evaporite deposition can be adduced.   An important observation is that restricted circulation combined with a high rate of net evaporation (evaporation minus rainfall) is a necessary prerequisite for supersaline brine formation from sea water.   Exchange of water with the open ocean, if too rapid, would prevent the buildup of dissolved salts by evaporation because of the large water capacity of the ocean.   Restricted circulation is accomplished by barrier sand bars, reefs, or tectonic sills which help to isolate bodies of evaporating brine from the open ocean.

Another important observation is the common occurrence of thick, areally extensive, halite-free, marine anhydrite beds.   In order to produce a monomineralic bed of this type, the degree of evaporation of sea water must remain relatively constant so that the water does not become saturated with halite.   Simple evaporation of sea water in a closed basin cannot be invoked. Over 2000 meters of sea water must be evaporated to produce just one meter of anhydrite before halite saturation is reached.   Thick monomineralic beds, therefore, must form from an open system.   To maintain a constant brine concentration and still permit continued $CaSO_4$ precipitation, a circulation system of the type shown in Fig. 5-4 must be maintained.   This is a constant-volume process where water lost through evaporation is replenished by inflow into the evaporite basin from the open ocean (King, 1947).   NaCl concentra-

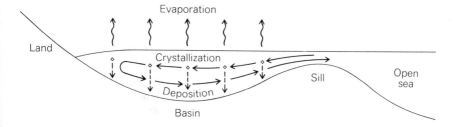

Fig. 5-4 Simplified open system model of an evaporite basin.   (*Adapted from King, 1947,* and others.)   Flow directions are indicated by solid arrows.   Note inward sloping water surface due to evaporation.

tion is held constant by a subsurface return of concentrated dense brine to the sea. Mathematically this open system model can be expressed in the form

$$\frac{dn_{H_2O}}{dt} = 55.5(R_{rain} + R_{rivers} - R_{evap}) + R_{SW}\, c_{H_2O}^{SW} - R_{brine}\, c_{H_2O}^{brine} \tag{22}$$

$$\frac{dn_{NaCl}}{dt} = R_{SW}\, c_{NaCl}^{SW} - R_{brine}\, c_{NaCl}^{brine} \tag{23}$$

$$\frac{dn_{CaSO_4}}{dt} = R_{SW}\, c_{CaSO_4}^{SW} - R_{brine}\, c_{CaSO_4}^{brine} - G \tag{24}$$

where n = total mass in moles in the basin

$\quad\quad t$ = time

$\quad\quad c$ = concentration in moles per liter

$\quad\quad R$ = volume flux to and from sources and sinks due to each process

$\quad\quad G$ = rate of deposition of $CaSO_4$

$\quad$ SW = sea water

At steady state the three expressions above can be set equal to zero and then solved for the rate of net evaporation:

$$\bar{R} = G\,\frac{(c_{NaCl}^{brine}/c_{NaCl}^{SW})\, c_{H_2O}^{SW} - c_{H_2O}^{brine}}{(c_{NaCl}^{brine}/c_{NaCl}^{SW})\, c_{CaSO_4}^{SW} - c_{CaSO_4}^{brine}} \tag{25}$$

where $\bar{R} = 55.5[R_{evap} - (R_{rain} + R_{rivers})]$ = net evaporation. Thus, if reasonable guesses for values of c can be made, the rate of net evaporation can be calculated from the rate of deposition or vice versa. For example assume evaporation to one-tenth the original volume so that

$\quad\quad c_{H_2O}^{SW} = 55 \quad\quad\quad\quad c_{H_2O}^{brine} = 50$

$\quad\quad c_{NaCl}^{SW} = 0.5 \quad\quad\quad\quad c_{NaCl}^{brine} = 5$

$\quad\quad c_{CaSO_4}^{SW} = 0.010 \quad\quad\quad c_{CaSO_4}^{brine} = 0.015$

Then, from Eq. (25):

$$\bar{R} = 6000G \tag{26}$$

The units for $\bar{R}$ and for G are moles per unit time. Dividing each side by the area of the basin floor and converting moles to cm^3 for water and to grams for $CaSO_4$ (assuming deposition as anhydrite)

$$\bar{R}^* = 800G^* \tag{27},$$

where $\bar{R}^*$ = rate of net evaporation, cm/yr

$\quad\quad G^*$ = rate of deposition of anhydrite, $g\ cm^{-2}\ year^{-1}$

Equation (27) can be applied to the formation of anhydrite in the Permian Castile formation of the southwestern United States, because the presence of subordinate halite suggests evaporation to about one-tenth original volume on the average. The rate of deposition obtained by Udden (1924) from an average thickness of presumably annual varves in the Castile formation is 0.16 cm/year of sediment consisting of about 70 percent anhydrite and 30 percent dolomite. This corresponds to 0.32 g anhydrite cm^{-2} $year^{-1}$. From Eq. (27) this value yields

$$\bar{R}^* = 260 \text{ cm/year}$$

By a similar calculation but based on the total volume of Castile evaporites (including nonvarved sediments) King (1947) calculated an annual evaporation rate of 285 cm/year. Rates of evaporation on this order are very high compared to average values for the earth but are not unreasonable. For very hot and arid climates net evaporation may exceed 250 cm/year (Langbein, 1961).

More elaborate evaporite basin models have been proposed to account for the simultaneous precipitation of two or more minerals at different basin locations, the result being the development of evaporite facies (Scruton, 1953; Briggs, 1958; Briggs and Pollack, 1967). In the former two papers a model similar to Fig. 5-4 is used, where density increases away from the inlet and precipitation of halite occurs in distal parts of the inflow water "tongue" while gypsum and carbonates precipitate closer to the inlet. In the latter paper, fluid-dynamical calculations led to the conclusion that the accumulation of the greatest percentage and thickness of halite in the center of a circular basin, such as the Michigan Basin, cannot be brought about by inflow from an inlet on the periphery of the basin. Radial inflow from many directions is necessary to produce a concentric thickness pattern.

Recently Schmalz (1969) has emphasized the importance of evaporite precipitation in deep-silled basins and has proposed a model for deep-water evaporite formation. According to Schmalz's model thick evaporite sequences would be explained simply by the filling of an originally deep basin with precipitated salts. Otherwise very rapid subsidence would be required to account for the rapid accumulation of thick salt sequences if they formed only in shallow water such as in salinas. It is probable, however, that due to an added mass of salt some subcrustal isostatic adjustment is made, resulting in limited subsidence. In this case the thickness of the salt formation could be greater than the water depth in the original basin, but it is still reasonable to assume some filling of the basin. Tests of shallow-water- versus deep-water-basin models must await further evidence for ancient water depths. Unfortunately, recent sediments shed little light on the problem because there are no known aerial extensive, well-developed, modern marine evaporite basins.

Evaporite minerals need not form only from a large standing body of water either shallow or deep. Kinsman (1969) and others have shown that evaporites also form upon and within extensive supratidal flats called sabkhas. Modern examples are the Trucial Coast of Arabia on the Persian Gulf and the northwestern margin of the Gulf of California. Marine sabkhas result from the evaporation of sea water brought in with infrequent high storm tides upon otherwise subaerial mud and sand flats. The sea water is occasionally left standing in very shallow "pans," but more generally recedes into the sediments to form saline ground water. In very arid climates the ground water and brine pans may undergo sufficiently rapid evaporation so that extensive precipitation of gypsum and halite occurs before the next influx of sea water. This is especially true of the pans which, due to a lack of occluding sediment grains, undergo more rapid evaporation. Incoming tides may redissolve halite so that it often does not accumulate, but gypsum generally remains. Some of the gypsum is subsequently replaced by anhydrite. Interstitial brines may also be supplied by subsurface inflow from the sea in regions where the ground-water table slopes inland due to evaporation (note the similarity here to the model evaporite basin (Fig. 5-4).

In the Persian Gulf sabkhas, evaporative concentration is accompanied by the dolomitization of fine-grained calcium carbonate in the sediments (see Chapter 8). This results in the formation of additional gypsum due to the generation of calcium ions by the dolomitization reaction:

$$Mg^{++} + 2CaCO_3 \rightarrow CaMg(CO_3)_2 + Ca^{++}$$

Thus, three kinds of gypsum can be found in the sabkha environment: that in brine pans formed by the evaporation of sea water, that in sediments formed by the evaporation of ground water, and that in $CaCO_3$-rich sediments formed as a result of dolomitization. The former two types can be distinguished by crystal morphology. Gypsum crystals formed within the sediments are stubby and cloudy due to numerous inclusions of the grains of the host sediment whereas crystals formed in the brine pans are elongated and essentially clear. A more elongate shape for the pan-derived material may be due to faster crystallization resulting from more rapid evaporation (Kinsman, 1969).

REFERENCES

Briggs, L. I., 1958, Evaporite facies, *Jour. Sed. Petrology*, v. 28, pp. 46–56.
———, and Pollack, H. N., 1967, Digital model of evaporite sedimentation, *Science*, v. 155, pp. 453–456.
Dellwig, L. F., 1955, Origin of the Salina salt of Michigan, *Jour. Sed. Petrology*, v. 25, pp. 83–110.
———, and Evans, R., 1969, Depositional processes in Salina salt of Michigan, Ohio, and New York, *Am. Assoc. Petroleum Geologists Bull.*, v. 53, pp. 949–956.
Hardie, L. A., 1967, The gypsum-anhydrite equilibrium at one atmosphere pressure, *Am. Mineralogist*, v. 52, pp. 171–200.

King, R. H., 1947, Sedimentation in Permian Castile Sea, *Am. Assoc. Petroleum Geologists Bull.*, v. 31, pp. 470–477.

Kinsman, D. J. J., 1966, Gypsum and anhydrite of recent age, Trucial Coast, Persian Gulf, in *Second symposium on salt, Cleveland, Ohio, Northern Ohio Geological Soc.*, v. 1, pp. 302–306.

———, 1969, Modes of formation, sedimentary associations, and diagnostic features of shallow-water and supratidal evaporites, *Am. Assoc. Petroleum Geologists Bull.*, v. 53, pp. 830–840.

Langbein, W. B., 1961, The salinity and hydrology of closed lakes, *U.S. Geol. Survey Prof. Paper* 412, 20 p.

MacDonald, G. J. F., 1953, Anhydrite-gypsum equilibrium relations, *Am. Jour. Sci.*, v. 251, pp. 884–898.

Phleger, F. B., 1969, A modern evaporite deposit in Mexico, *Am. Assoc. Petroleum Geologists Bull.*, v. 53, pp. 824–829.

Posnjak, E., 1940, Deposition of calcium sulfate from sea water, *Am. Jour. Sci.*, v. 238, pp. 559–568.

Robinson, R. A., and Stokes, R. H., 1959, *Electrolyte solutions*, New York, Academic, 2d ed., 559 p.

Schmalz, R. F., 1969, Deep-water evaporite deposition: A genetic model, *Am. Assoc. Petroleum Geologists Bull.*, v. 53, pp. 798–823.

Scruton, P. C., 1953, Deposition of evaporites, *Am. Assoc. Petroleum Geologists Bull.*, v. 37, pp. 2498–2512.

Udden, J. A., 1924, Laminated anhydrite in Texas, *Geol. Soc. Amer. Bull.*, v. 35, pp. 347–354.

Usiglio, J., 1849, Analyse de l'eau de la Mediterranée sur les côtes de France, *Annalen der Chemie*, v. 27, pp. 92–107, 172–191.

6
Diagenetic Processes

The purpose of this chapter is to demonstrate how the general problem of diagenesis can be approached from an analytical standpoint and to discuss in detail several important diagenetic processes which are mentioned only incidentally in later chapters. These processes include compaction, cementation, diffusion, mineral segregation, and phenomena resulting from the existence of the Donnan equilibrium.

THE DIAGENETIC EQUATION

Any given property of a sediment or sedimentary rock, such as mineralogy, fossil content, pore water composition, etc., can be expressed as a function of spatial position and time:

$$p = f(x,y,z,t) \tag{1}$$

where p = sediment property

x = vertical coordinate

y, z = horizontal coordinates

t = time

Diagenesis, as used in this book, refers to changes that take place within a sediment during and after burial.* Consequently, for the description of diagenesis, especially early diagenesis, the convention is adopted that x in Eq. (1) represents depth below the sediment-water interface, measured positively downward, and t represents time since deposition. Also, since most of the diagenetic changes described in this book are of a predominantly vertical nature (e.g., compaction), lateral variations, except in the special case of concretion formation, will be neglected. This means that the variables y and z can be neglected, and if Eq. (1) is differentiable,

$$dp = \frac{\partial p}{\partial x}\bigg|_{t} dx + \frac{\partial p}{\partial t}\bigg|_{x} dt \qquad (2)$$

Dividing each side by dt and rearranging,

$$\frac{\partial p}{\partial t}\bigg|_{x} = \frac{dp}{dt} - \frac{\partial p}{\partial x}\bigg|_{t} \omega \qquad (3)$$

where $\omega = dx/dt$. Here ω is the "rate of deposition" as measured by radiometric dating. It is actually the true rate of deposition minus the sum of the rates of erosion and compaction. Equation (3) is designated as the *general diagenetic equation*.

The total derivative dp/dt in Eq. (3) includes all diagenetic processes affecting p. In the special or degenerate case where there are no diagenetic changes, $dp/dt = 0$, and all variations in p with depth are due to variations in p at the time of deposition. As a result

$$\frac{\partial p}{\partial t}\bigg|_{x} = -\omega \frac{\partial p}{\partial x}\bigg|_{t} \qquad \text{(no diagenesis)} \qquad (4)$$

An example of this situation is where p represents the median size of sand grains, which can vary with time as a result of a fluctuating degree of turbulence in water overlying the site of deposition.

The opposite situation is that of steady-state diagenesis. In steady-state diagenesis all processes are adjusted so that the property p does not change with time at any depth, i.e.,

$$\frac{\partial p}{\partial t}\bigg|_{x} = 0 \qquad \text{(steady-state diagenesis)} \qquad (5)$$

This means that all variations in p are due to diagenesis, and a layer at depth x, at the time it was deposited, had the same value for p as the layer presently being deposited. It also means that the shape of the curve of p versus x, relative to the sediment-water interface, does not change during deposition. This is illustrated in Fig. 6-1. The situation of steady-state diagenesis,

---

* Rapid changes that occur prior to burial are referred to as halmyrolysis (see Chapter 9).

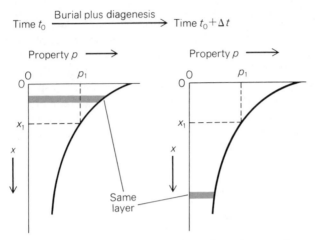

Fig. 6-1 Diagrammatic representation of steady-state diagenesis. Note that at a given *depth* x_1 the sediment property $p = p_1$ does not change with time but that p for a given *layer* changes as it is buried.

although never strictly correct for any given sediment, is still a very useful concept and is employed frequently in this book. It serves as an idealized model, for those changes attributable solely to diagenesis, with which the measured properties of real sediments can be compared.

COMPACTION

POROSITY AND COMPACTION

Loss of water from a sediment, due to compression arising from the deposition of overlying sediment, is here defined as *compaction*. The most useful measure of water content for a sediment, whose pores are filled with water, is the porosity, which is defined as

$$\Phi = \frac{V_{\text{water}}}{V_{\text{water}} + V_{\text{solids}}} \tag{6}$$

where V = volume

Φ = porosity

Original porosity of a sediment at the time of deposition is a function primarily of mineralogy. Mineralogy is important because of the distinctly different mode of packing of clay minerals as compared to other common minerals such as quartz, feldspar, etc. Clay minerals occur as very fine (most <2 microns) platelets which pack together in a loose fashion due to surface electrostatic phenomena. Because of their fine grain size and consequent large specific surface area, electrostatic interparticle forces are strong and help to maintain an open structure (Engelhardt, 1960). Some flocculated clays form a "house of cards" structure consisting of platelets joined in edge-to-face or edge-to-edge fashion. Face-to-face dense packing is opposed because of repulsion between similarly charged double layers on the flat surfaces. Different clay minerals exhibit different packing because of differences in the nature of their electrical double layers, and thus exhibit different porosities (see Fig. 6-5). The closeness of packing is also affected by the salinity and chemical composition of the water contacting clay minerals, but the relationship is complex and no simple generalization can be made (Meade, 1966). As a result of electrostatic open packing, the initial porosity of fine-grained sediments containing a large proportion of clay minerals is high (see Fig. 6-2).

In contrast to clays, quartz and feldspar, as well as most other minerals, pack together in a more or less simple geometric fashion. This is because electrostatic effects are negligible due to much coarser grain sizes (little quartz or feldspar occurs at sizes less than 2 microns). For well-sorted, subspherical sand grains extremes of porosity can be calculated theoretically from the various packing geometries for equal-sized spheres. The range is $\Phi = 0.26$ to 0.48. By comparison, measurements for surficial well-sorted sands range from 0.36 to 0.46 (Engelhardt, 1960). Increase in angularity causes a slight increase in initial porosity, whereas poor sorting causes a large decrease in porosity. Poor sorting is due to the filling of the interstices between sand grains with finer sand, silt, and clay. Also, mica content of a sand can affect its porosity, especially during compaction, because of its flat shape and ability to wrap around grains and fill interstices. A comparison of porosities of some natural surficial sands, silts, and clays is shown diagrammatically in Fig. 6-2.

Since sand grains at the time of deposition are already packed in a more or less tight coordinating geometry, little change in the porosity of sands occurs

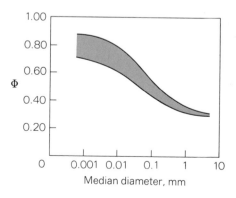

Fig. 6-2 Relation between porosity Φ and median particle diameter in terrigenous (noncarbonate) recent, surface or near-surface sediments. Note the increase in Φ with decrease in size, reflecting an increase in percentage of clay minerals. (*After Meade, 1966.*)

as a result of burial.   For very angular sands some reorientation takes place, and compacted angular sands become slightly less porous than equivalent well-rounded sands (Meade, 1966).   Only when great depths are reached, and the grains begin to interpenetrate as a result of pressure solution (see Fig. 6-6) does porosity change appreciably due to pressure.   Porosity of sands also may decrease at any time by the precipitation of interstitial cement from pore waters.

Fine-grained sediments rich in clay minerals, in contrast to sands, undergo continuous compaction during burial due to the continuous increase in pressure of the overburden.   This is shown in Figs. 6-3 and 6-4.   The open house of cards structure begins to be compressed as pressure opposes the electrostatic forces holding the "cards" apart.   The rate of compression depends upon the rate at which water can be expelled.   This in turn depends upon the rate of deposition, i.e., rate of pressure increase, and the permeability of the sediments.   Differing degrees of compaction for different clay minerals at the same rate of pressure loading are illustrated by the experimental results shown in Fig. 6-5.   The differences, for montmorillonite and kaolinite, can be explained at least partly on the basis of different permeabilities.   Montmorillonite, which adsorbs much more water than kaolinite (see Chapter 9), is less permeable (Bredehoeft and Hanshaw, 1968) and compacts less readily.

The effect of different rates of deposition is shown in Fig. 6-3.   The sediment from Lake Mead is less compacted than the much more slowly deposited muds from the basins off southern California.   Although the sediments vary somewhat in clay content and clay mineralogy there is little doubt that the Lake Mead sediments owe their high porosity at least partly to a very high rate of deposition.   If deposition is sufficiently rapid over a long enough period of time, the expulsion of water may not keep pace with pressure loading, and consequently the pore waters may begin to support an appreciable

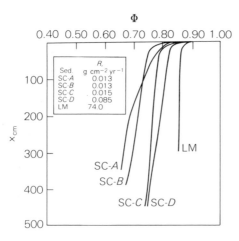

Fig. 6-3 Porosity, Φ, versus depth for fine-grained modern sediments from marine basins off southern California and Lake Mead.   LM = Lake Mead, SC = southern California. R = rate of deposition.   (*After Emery, 1960.*)

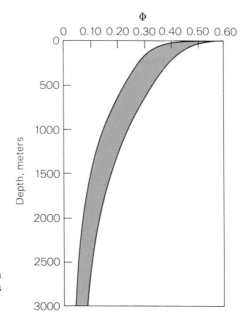

Fig. 6-4 Plot of porosity Φ versus depth for some fine-grained marine terrigenous sediments. (*After Hedberg, 1936.*)

proportion of the weight of the overlying sediment particles. As a result anomalously high porosities are encountered at depth, and pressures in the pores may considerably exceed those expected for simple hydrostatic burial. Excess pore water pressure in deeply buried sediments from the Gulf Coast (Dickinson, 1953) can be explained by this mechanism. Bredehoeft and Hanshaw (1968) have shown theoretically that for an average rate of deposition (minus compaction) of 0.05 cm/year, excess pore pressure approaching lithostatic pressure, i.e., the situation where the pore waters bear the entire weight

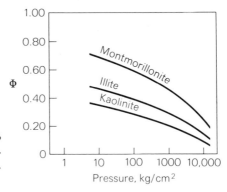

Fig. 6-5 Relation between porosity Φ and pressure for experimentally compacted samples of different clay minerals. (*After Meade, 1966.*)

of the overlying particles, can be achieved in the Gulf Coast sediments providing that they have permeabilities similar to values determined in the laboratory on clays and shales.

During deep burial, factors other than compression of the house of cards structure become important. According to Hedberg (1936) and Weller (1959) once a low enough porosity is reached, fine-grained sediments begin to undergo a different type of compactive process. This can be seen in Fig. 6-4 by the change in slope at $\Phi \approx 0.35$. Below this porosity the rate of compaction is assumed to be controlled by the deformation and filling by clays of interstices between larger silt and sand grains. Below about $\Phi = 0.10$ crushing of larger grains and recrystallization of clay is believed to be important in eliminating the remaining pore space. By this time the sediment has been buried to several thousand meters.

COMPACTION CALCULATIONS FOR SURFICIAL SEDIMENTS

Several useful parameters can be obtained from Φ versus x plots for surficial fine-grained sediments, such as those shown in Fig. 6-3 (Emery and Rittenberg, 1952). For instance, for diagenetic chemical calculations it is useful to know how much water has passed a given horizon since original burial, and for measurements of rate of deposition, corrections for the degree of compaction at each depth should be made. In the following discussion equations enabling these types of calculations will be derived. The reason why the discussion is confined to near-surface (<10 meters depth) sediments is because, for several derivations, steady-state diagenesis is assumed and this situation is approached most reasonably when shorter sections of sediment are considered. Porosity in near-surface fine-grained sediments is usually calculated from weight percent water as measured by loss of weight upon drying at 100°C. The conversion equation is

$$\Phi = \frac{Wd_s}{Wd_s + (1 - W)d_w} \tag{7}$$

where W = weight percent water (wet weight)/100

d_s = average density of sediment particles

d_w = density of pore water

From the rate of deposition and porosity versus depth curves five useful parameters can be calculated.

1. *Thickness of an annual sediment layer.* As a result of compaction the thickness representing a given interval of time decreases with depth. The

volume of solids per unit area $\bar{V}_s$ within any sediment layer is related to the rate of deposition R in units of mass/unit area/year by

$$\bar{V}_s = \frac{R \Delta t}{d_s} \tag{8}$$

where Δt = time in years to deposit the volume, $\bar{V}_s$ per unit area. Now, the volume of total sediment (water plus solids) per unit area, within an annual layer, is simply the thickness of an annual layer $\Delta x/\Delta t$, which is denoted as ω.

Therefore, from Eqs. (6) and (8)

$$1 - \Phi = \frac{\bar{V}_s}{\bar{V}_s + \bar{V}_w} = \frac{R \Delta t}{d_s \omega \Delta t} \tag{9}$$

Rearranging

$$\omega = \frac{R}{d_s} \left(\frac{1}{1 - \Phi} \right) \tag{10}$$

If R remains constant over a considerable thickness of deposition, then Eq. (10) can be used to calculate ω at any given depth from the curve of Φ versus x.

2. *Total compaction.* The total compaction since deposition of a sediment layer of thickness h can be expressed as $(h_{t_0} - h)/h_{t_0}$, where h_{t_0} refers to h at the time of deposition. During compaction, $\bar{V}_s$ for a given layer remains constant. Thus

$$\bar{V}_s = \bar{V}_{s_{t_0}} \tag{11}$$

and from Eq. (6)

$$(1 - \Phi)h = (1 - \Phi_{t_0})h_{t_0} \tag{12}$$

Rearranging

$$h_{t_0} = \left(\frac{1 - \Phi}{1 - \Phi_{t_0}} \right) h \tag{13}$$

therefore

$$\frac{h_{t_0} - h}{h_{t_0}} = \frac{\Phi_{t_0} - \Phi}{1 - \Phi} \tag{14}$$

If steady-state diagenesis has existed over the period since t_0, then

$$\Phi_{t_0} = \Phi_0 \tag{15}$$

where $\Phi_0 = \Phi$, presently at $x = 0$. Finally, if compaction is expressed as $\Delta h/h_0$, for steady-state diagenesis

$$\frac{\Delta h}{h_0} = \frac{\Phi_0 - \Phi}{1 - \Phi} \tag{16}$$

3. *Rate of compaction.* Rate of compaction can be expressed as $d\Phi/dt$. If $\Phi = f(x,t)$ then from the general diagenetic equation (3)

$$\frac{d\Phi}{dt} = \frac{\partial \Phi}{\partial t}\bigg|_x + \frac{\partial \Phi}{\partial x}\bigg|_t \omega \tag{17}$$

If steady-state diagenesis exists, then $\dfrac{\partial \Phi}{\partial t}\bigg|_x = 0$ and

$$\frac{d\Phi}{dt} = \frac{\partial \Phi}{\partial x}\bigg|_t \omega \tag{18}$$

or

$$\frac{d\Phi}{dt} = \frac{\partial \Phi}{\partial x}\bigg|_t \frac{R}{d_s}\left(\frac{1}{1 - \Phi}\right) \tag{19}$$

For a given depth, $\dfrac{\partial \Phi}{\partial x}\bigg|_t$ and Φ can be obtained from Φ versus x curves.

4. *Rate of water flow through a horizon.* Let Q be the upward flow of water due to compaction in units of volume of water per unit area of sediment per year. This flow is equal to the total volume of upward flowing water below a horizon at time t minus the total volume beneath the same horizon at t plus one year. Mathematically

$$Q = \left[\int_x^X \Phi(x)\,dx\right]_t - \left[\int_{x+\omega}^{X'} \Phi'(x)\,dx\right]_{t+1} \tag{20}$$

where $t + 1$ represents 1 year later and X and X' represent the depth to where continuous upward flow is interrupted. If steady-state diagenesis exists

$$\Phi(x) = \Phi'(x) \tag{21}$$

$$X' = X + \omega_X$$

where $\omega_X = \omega$ at $x = X$

so that

$$Q = \int_x^{x+\omega} \Phi(x)\,dx - \int_X^{(X+\omega_X)} \Phi(x)\,dx \tag{22}$$

If an annual layer is thin, which is almost always true, Φ can be treated as a constant in the interval x to $(x + \omega)$. Thus, for steady state

$$Q = \Phi\omega - \Phi_x\,\omega_x = \frac{R}{d_s}\left(\frac{\Phi}{1 - \Phi} - \frac{\Phi_x}{1 - \Phi_x}\right) \tag{23}$$

5. *Total water which has passed through a layer since deposition.* The volume of water per unit area of sediment, W_T, which has passed through a given layer at depth x since deposition is equal to the total volume of upward flowing water beneath the layer at the time of deposition minus the total volume presently beneath it. Mathematically this can be expressed by analogy with Eq. (20) as

$$W_t = \left[\int_0^X \Phi(x)\,dx\right]_{t_0} - \left[\int_x^{X'} \Phi'(x)\,dx\right]_{t=\text{now}} \tag{24}$$

If steady-state diagenesis exists

$$\Phi(x) = \Phi'(x)$$

$$X' = X + x'$$

where $x' =$ thickness of sediment below X representing the same time interval as x. Thus, for steady state

$$W_t = \int_0^x \Phi(x)\,dx - \int_x^{X+x'} \Phi(x)\,dx \tag{25}$$

CEMENTATION

The filling of the pores of a sediment with precipitated mineral matter during diagenesis is here defined as *cementation*. Along with compaction, cementation is one of the two major processes whereby loose sediment is converted into hard sedimentary rock. Because cements are much more easily studied in sandstones than in shales, only the cementation of sand will be discussed in this section. Also, because of their quantitative predominance, discussion will be confined to calcium carbonate and quartz cements.

Cementation in sands takes place as a result of the contact of supersaturated pore waters with crystallization nuclei which, at least initially, are sand grains. Because of the abundance of grain surfaces per unit volume of pore solution, heterogeneous nucleation is promoted and supersaturations necessary for precipitation thereby are relatively low. At shallow burial depths, precipitation at grain boundaries is especially favored because of the presence of small interstices and cracks (Wollast, 1969). For a very small crack with dimensions on the order of the size of the critical nucleus, the free energy of nucleation can be given by an expression analogous to Eq. (75) of Chapter 2:

$$\Delta F^* = -n^* \oplus + (\sigma_s - \sigma_n) A \qquad (26)$$

where σ_s = specific interfacial free energy between the nucleus and host grains

σ_n = specific interfacial (surface) free energy between host grains and water

A = surface area of the crack

Since σ_s is generally considerably less than σ_n, the surface-energy term $(\sigma_s - \sigma_n) A$ is negative, and precipitation can take place in a very small crack *even from an unsaturated solution.*

As a sediment is buried, cementation at grain boundaries eventually becomes less favorable. This is because a competing process, pressure solution, begins to be important. Pressure solution results from the appreciable excess pressure generated at horizontal grain contacts when sand is buried to considerable depth. Because solubility increases with pressure, a decreasing concentration gradient between the grain contact zone and the adjacent lower pressure pore space is built up. In response to the concentration gradient outward diffusion takes place and the particles dissolve in the contact zone. Quantitative theoretical treatment of this problem has been presented by Weyl (1959). The result of dissolution is that the adjacent sand grains interpenetrate one another to produce sutured contacts (Fig. 6-6) and the porosity is accordingly reduced. [Sippel (1968) has recently suggested that many sutured grain contacts also arise from the growth interference of secondary overgrowths on

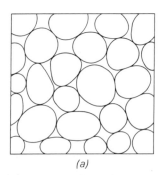
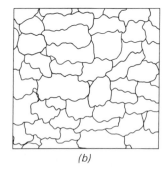

(a)   (b)

Fig. 6-6 Diagrammatic representation of the pressure solution of sand grains. (*a*) Cross section of sand grains before pressure solution; (*b*) cross section of sand grains after pressure solution. Note interpenetrating, sutured contacts.

detrital quartz grains; however, some of the overgrowths may have been originally present on the grains during sedimentation and the suturing brought about by the pressure solution of the overgrowths.]

The silica released by pressure solution of quartz sand grains is probably a major source of silica for the cementation of sandstones. It may be precipitated in pore spaces adjacent to where pressure solution takes place or may migrate to cement more distant particles. The common occurrence of silica cementation without evidence of accompanying pressure solution (Siever, 1959) suggests that migration of silica is important.

Not all deeply buried quartz grains undergo pressure solution. Heald (1955) and Weyl (1959) have noted that pressure solution is most effective where clay films are located at grain contacts. Weyl has emphasized the importance of clay as a medium for enhancing diffusion of dissolved silica. The enhanced diffusion is due to the much larger number of water films provided by the adsorbed and interlayer water in a moderate thickness of clay, as compared to the single film which is present at the grain contact in the absence of clay. Apparently, lower diffusion coefficients in the adsorbed and interlayer water are compensated by the greater cross section for diffusion.

Cementation by silica must be predominantly a phenomenon of later diagenesis because almost no examples are found in recent marine sediments. By contrast, cementation by calcium carbonate may occur rapidly after deposition. A good example is beachrock. Beachrock is beach and intertidal sand (usually carbonate and skeletal fragments) which is cemented by $CaCO_3$ in subtropical to tropical climatic zones. The cement forms so rapidly that human artifacts only a few tens of years old are commonly found cemented into the beachrock. Other examples of early diagenetic $CaCO_3$ cementation are provided by subtidal lithified calcarenites, reefs, and pelagic oozes (for a summary of numerous occurrences see Mackenzie et al., 1969). In all cases the cements are aragonite or high-magnesium calcite indicating formation from sea water or other magnesium-rich solutions (see Chapter 8). The cement in all these occurrences appears to be derived from external sources such as dissolved Ca^{++} and HCO_3^- in sea water; consequently, constant flushing of the sediments by fresh, supersaturated water is necessary to enable the buildup of appreciable quantities of cement. The need for constant flushing explains why cementation usually occurs in sands rather than muds. The reason why some recent sands are cemented while many others are not has not yet been satisfactorily explained.

Cementation by $CaCO_3$ during later diagenesis must also be important. The sources of dissolved Ca^{++} and HCO_3^- may in many cases be original $CaCO_3$ dissolved by pressure solution during stylolite formation. Also, near-surface cementation of marine limestone by meteoric waters after uplift above sea level is common and has been studied rather extensively (see Friedman, 1964). It appears to take place rapidly. The cement is locally derived by the dissolution of grains of unstable carbonate minerals (aragonite, high-magnesium calcite) and precipitated as replacements or void fillings consisting of low-magnesium calcite.

DIFFUSION IN SEDIMENTS

THE DIFFUSION EQUATIONS

In near-surface subaqueous sediments, changes in the composition of pore waters with depth are generally much greater than lateral changes.   Therefore, vertical concentration gradients predominate and one-dimensional diffusion equations, i.e., Fick's first and second laws of diffusion, are applicable, with the unidimensional coordinate x being depth.   Before applying Eqs. (63) and (64) of Chapter 2 directly to sediments, modification must be made, because in a sediment, free diffusion is hindered by collision and interaction of diffusing species with sediment particles.   Retardation of diffusion is commonly expressed in terms of the porosity of the sediment Φ, and the tortuosity θ. Tortuosity is defined as

$$\theta = \frac{dl}{dx} \tag{27}$$

where dl = length of the actual sinuous diffusion path over a depth interval dx. Since $l \geq x$, $\theta \geq 1$.   In terms of porosity and tortuosity, Eqs. (63) and (64) of Chapter 2 become

$$J_x = \frac{-D\Phi}{\theta^2} \frac{\partial C}{\partial x} \tag{28}$$

where C = concentration in mass per unit volume of pore water

$\quad\quad J_x$ = diffusional flux in mass/area of sediment/time

$\quad\quad D$ = Fick's law diffusion coefficient

and

$$\frac{\partial C}{\partial t} = -\frac{1}{\Phi} \frac{\partial J_x}{\partial x}$$

$$= \frac{D}{\theta^2} \left[\frac{\partial^2 C}{\partial x^2} + \left(\frac{1}{\Phi} \frac{\partial \Phi}{\partial x} - \frac{2}{\theta} \frac{\partial \theta}{\partial x} \right) \frac{\partial C}{\partial x} \right] \tag{29}$$

For most natural sediments the change of θ with depth is not known.   Also, $\partial \Phi / \partial x$ below the top few centimeters is relatively small.   Therefore, as a very crude first approximation the second term within the brackets of Eq. (29) can be neglected giving the simplified equation

$$\frac{\partial C}{\partial t} = D_s \frac{\partial^2 C}{\partial x^2} \tag{30}$$

where $D_s = D/\theta^2 =$ whole sediment diffusion coefficient. The value of D, the simple Fick's law diffusion coefficient, for most dissolved salts at room temperature is in the range 1 to 2×10^{-5} cm²/sec. Therefore, since $\theta \geqq 1$, $D_s \leqq 2 \times 10^{-5}$ cm²/sec. Measured values of D_s, however, can be much less than that predicted by D/θ^2. This is mainly true of cations in clay-rich fine-grained sediments. Due to cation exchange with negatively charged clays, cation mobility is hindered relative to anions, which undergo little or no exchange with clays. A comparison of cation and anion diffusion in bentonite (montmorillonite) clay-water gels has been demonstrated by the work of Lai and Mortland (1962). Table 6-1 is taken from their work. Note that the diffusion of sulfate is approximately the same as in pure water demonstrating a low tortuosity in the experimental gels (as might be expected since they contain about 90 percent water). By comparison the rate of diffusion of Na⁺ is retarded because of the time spent on clay surface exchange sites before "hopping" to adjacent sites.

Under certain circumstances the effect of cation exchange upon diffusion coefficients can be calculated. This has been shown by Duursma and Hoede (1967). If exchange involves cations of the same valence which form an ideal solution on the clay exchange sites, then for exchange equilibrium (assuming $\gamma_{A^+} = \gamma_{B^+}$)

$$A^{+m} + B\text{-clay} \leftrightharpoons B^{+m} + A\text{-clay}$$

$$K = \frac{C_B \bar{C}_A}{C_A \bar{C}_B} \tag{31}$$

where $C =$ concentration of each dissolved ion per unit volume of interstitial water

$\bar{C} =$ concentration of each ion on the clay sites per unit volume of interstitial water

Table 6-1   Diffusion of sodium and sulfate ions in a 10 percent suspension of Na bentonite in water at 25°C.   [After Lai and Mortland (1962)]

Ion	$D(cm^2/sec)$ in clay	$D(cm^2/sec)$ in water
Na⁺	0.54×10^{-5} 0.47×10^{-5}	1.35×10^{-5}
SO₄⁻⁻	1.15×10^{-5} 1.25×10^{-5}	1.1×10^{-5}

If the diffusing ion A^{+m} displaces only a very small proportion of the ions originally on the clay, then the ratio $C_B/\bar{C}_B$ remains essentially constant during diffusion. For this situation

$$\frac{\bar{C}_A}{C_A} = \frac{K\bar{C}_B}{C_B} = \text{const} \tag{32}$$

Now if local ion exchange equilibrium is rapidly maintained during diffusion, the concentration of A^{+m} in solution is affected both by diffusion and by loss or gain from exchange sites, so that

$$\frac{\partial C_A}{\partial t} = \frac{D}{\theta^2} \frac{\partial^2 C_A}{\partial x^2} - \frac{\partial \bar{C}_A}{\partial t} \tag{33}$$

Substituting Eqs. (32) in (33) and rearranging

$$\frac{\partial C_A}{\partial t} = \left(\frac{D/\theta^2}{1 + K\bar{C}_B/C_B} \right) \frac{\partial^2 C_A}{\partial x^2} \tag{34}$$

In this case the expression in parentheses can be identified as D_s, the empirical diffusion coefficient.

As an example, consider montmorillonitic marine sediment with total cation-exchange capacity of 30 milliequivalents per 100 grams dry weight and exchange sites 75 percent occupied by Mg^{++} (see Chapter 9). If the sediment contains 50 percent water by weight, the value of $\bar{C}_B/C_B$ for Mg^{++} is 2.3. If the diffusing ion is Ca^{++}, the value for K from Chapter 9 is approximately 2. Therefore, for Ca^{++},

$$D_s = \frac{0.18 D}{\theta^2}$$

From this it would be expected that the diffusion coefficient of Ca^{++} ion in marine montmorillonitic sediment would be considerably lower than D, the value for sediment-free sea water.

Factors other than exchange with clays can bring about lowered values for the diffusion coefficient of ions in sediments. These include increased viscosity of water near the sediment particles (Gast, 1963); anion exclusion in low-water content, highly compacted sediments (see page 110, Donnan phenomena); and the presence of dead end pore space (Goodknight et al., 1960). As a general rule one can assume that anions in near-surface clay-rich sediments have diffusion coefficients close to D/θ^2, whereas cations will have diffusion coefficients less than D/θ^2. In addition, doubly charged cations will generally diffuse slower, due to stronger interactions with clays, than singly charged cations, although specific exceptions exist; e.g., Cs_{aq}^+ diffuses more slowly than the common doubly charged cations.

SOLUTIONS OF THE DIFFUSION EQUATIONS

Solution of Eq. (28) is of major interest for calculating the flux between sediment and overlying water.   The correct equation is

$$J_x = -D_s \Phi \frac{\partial C}{\partial x}\bigg|_{x=0} \tag{35}$$

If the plot of C versus x can be described by an analytical expression, then the flux J_x is obtained by differentiating the expression and substituting $x = 0$. If the measured plot of C versus x is not readily described analytically, an approximate solution can be derived if the gradient in the upper few centimeters is roughly linear.   The solution of Eq. (35) for a linear gradient is

$$J_x = -D_s \Phi \frac{(C_{x'} - C_0)}{x'} \tag{36}$$

where $x' =$ a shallow depth within the linear gradient

$C_{x'} =$ concentration at $x = x'$

$C_0 =$ concentration at $x = 0$

Equation (36) is also an appropriate solution for the situation where a linear concentration gradient connects two reservoirs of constant concentration.   An example would be upward diffusion of Cl^- between a buried salt layer through homogeneous sediment to overlying sea water (Manheim and Bischoff, 1969).   The concentration of Cl_{aq}^- in the sea water is constant because of rapid convection (i.e., open system), while the concentration of Cl_{aq}^- is constant within the salt layer due to equilibrium with NaCl.   If the salt layer is not buried too deeply so that D_s is relatively constant over the distance of diffusion, then x' can be identified with depth to the salt layer and J_x calculated using Eq. (36).

Most diffusion calculations involve the solution of Eq. (30).   However, because diffusion is a response to chemical gradients generated by other processes, a complete description of factors affecting concentration often entails the inclusion of additional expressions for the generating mechanisms. In this book the generating mechanism is almost always a diagenetic chemical reaction so that diffusion is usually included only as a part of an overall process.   Thus, Eq. (30) by itself is rarely used.   Solution of Eq. (30) is most useful in describing the response of the dissolved ions in a sediment to changes in salinity of the overlying water during deposition.   For a brackish water environment, where salinity fluctuations can be large, calculation of the diffusional response in the underlying sediments has been made by Scholl and Johnson (1967).   Solutions of Eq. (30) for other types of boundary conditions can be found in Duursma and Hoede (1967) and Crank (1956) or through the use of heat flow analogues in Carslaw and Jaeger (1959).

DIFFUSION AND CHEMICAL REACTION

If diffusion in a sediment is the consequence of chemical reaction, the sum total of diagenetic processes affecting the concentration of a dissolved species can be divided into a diffusional term plus a chemical term.* Thus

$$\frac{dC}{dt} = D_s \frac{\partial^2 C}{\partial x^2} + \frac{dC}{dt} \text{ chem}$$

Substitution into the general diagenetic Eq. (3) gives

$$\left.\frac{\partial C}{\partial t}\right|_x = D_s \frac{\partial^2 C}{\partial x^2} + \frac{dC}{dt} \text{ chem} - \left.\frac{\partial C}{\partial x}\right|_t \frac{dx}{dt} \tag{37}$$

As an example of the solution of Eq. (37), assume that C represents the concentration of a dissolved species which is added to a sediment and then lost during diagenesis by first-order kinetics. Then, if steady-state diagenesis exists:

$$D_s \frac{\partial^2 C}{\partial x^2} - \omega \frac{\partial C}{\partial x} - k_1 C = 0 \tag{38}$$

where $\omega = dx/dt$

k_1 = first-order rate constant

If the boundary conditions are

$$x = 0 \qquad C = C_0$$

$$x \to \infty \qquad C \to 0$$

solution of Eq. (38) yields for constant ω

$$C = C_0 \exp\left[\left(\frac{\omega - (\omega^2 + 4k_1 D_s)^{\frac{1}{2}}}{2D_s}\right)x\right] \tag{39}$$

This simple model would be appropriate for the distribution of a dissolved species undergoing radioactive decay without precipitation or replenishment by dissolution of solids. Other applications of diagenetic chemical reaction-diffusion equations to specific problems are presented below and in later chapters.

MINERAL SEGREGATION

Due to diffusion, ground-water flow, and localized precipitation, chemical elements which are originally dispersed within a sediment or sedimentary rock

---

* Strictly speaking an additional term for the upward flow of water, due to compaction, should be included which, from Eqs. (10) and (23), is $(\omega - \omega_x)(\partial C/\partial x)$. For most sediment columns, however, interruptors of continuous upward flow (sand layers, basement rock) are sufficiently shallow that $\omega \approx \omega_x$ and, as a first approximation, the compactive flow term can be omitted.

can be concentrated into layers, concretions, nodules, etc., during diagenesis. One of the principal causes of this segregation is a heterogeneous distribution of decomposing organic matter which, as a result of bacterial activity (described in the next chapter), brings about the localized precipitation of minerals whose solubilities are affected by bacterial metabolites. Also important are scattered bodies originally in the sediment which induce precipitation because of their efficiency as seeding agents for heterogeneous nucleation. An example is a calcite shell fragment which can act as a nucleating agent for the formation of a calcite concretion. This section is not concerned with the formation of mineral concentrations in sediments which have resulted from the *in situ* recrystallization of originally deposited monomineralic bodies, such as recrystallized fossils or chert nodules derived from siliceous sponges. The major purpose is to discuss segregation which has resulted from transport processes, chiefly diffusion.

LAYER FORMATION

The formation of monomineralic layers by diffusion must result from strong vertical chemical gradients. Sharp vertical gradients are characteristic of the area adjoining the sediment-water interface and it is here where many monomineralic layers may form. An example is an anaerobic sediment, whose pore water is high in dissolved Mn^{++} or Fe^{++}, overlain by Mn^{++}- or Fe^{++}-free aerated water. Dissolved Mn^{++} and Fe^{++} are unstable in oxygenated water and are rapidly oxidized and precipitated by dissolved O_2. Strong concentration gradients between anaerobic sediment and aerobic water lead to diffusion of Fe^{++} and Mn^{++} to the oxygenated interface where precipitation takes place. If deposition of detrital material is slow relative to upward diffusion, as a result of continued upward diffusion and precipitation, moderately thick relatively monomineralic layers of iron oxide and/or manganese oxide can accumulate (see Fig. 6-7). This problem can be treated mathematically as follows: the

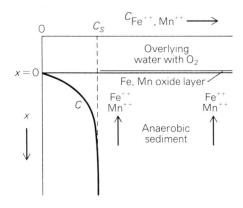

Fig. 6-7 Diffusion model for the formation of monomineralic Fe or Mn oxide layers. Arrows represent upward diffusion direction. The curve marked C represents a generalized plot of concentration of dissolved Fe^{++} or Mn^{++} with depth x in the sediment; C_S is the saturation concentration for the dissolving Fe or Mn containing particles. The surface marked $x = 0$ is the sediment-water interface.

processes governing Mn^{++} or Fe^{++} concentration in the sediment pore waters are solubilization of scattered minerals containing Mn or Fe and diffusion. If dissolution is diffusion-controlled, Eq. (100) of Chapter 2 is applicable. The general diagenetic equation then is

$$\left.\frac{\partial C}{\partial t}\right|_x = D_s \frac{\partial^2 C}{\partial x^2} + \frac{\bar{A}\Phi D_s(C_s - C)}{r} - \omega \frac{\partial C}{\partial x} \tag{40}$$

where $r =$ radius of Fe or Mn containing particles which are assumed to be of equal size and spherical

$\bar{A} =$ total surface area of dissolving particles per unit volume of interstitial water

$C_s =$ concentration of Fe^{++} or Mn^{++} at the surface of the dissolving particles

$\bar{A}$ can be expressed as

$$\bar{A} = \frac{3 M_F d_s}{r d_F}\left(\frac{1}{\Phi} - 1\right) \tag{41}$$

where $M_F =$ mass fraction of dissolvable iron or manganese per total mass of solids

$d_s =$ average density of total solids

$d_F =$ mass of Fe or Mn per unit volume of dissolving particles

Direct integration of Eq. (40), even for the situation of steady-state diagenesis, is exceedingly difficult because r is a function of time at each depth in the sediment. As dissolution proceeds, r goes to zero at successively greater depths; i.e., the source particles completely dissolve. Because of this, direct solution of Eq. (40) will not be attempted. Instead, the necessity of utilizing a model of this type for monomineralic layer formation will be demonstrated by considering the situation where there are no source particles and only original dissolved Fe^{++} and Mn^{++} in the pore waters are available for layer formation.

If the concentration of Fe^{++} or Mn^{++} in the interstitial waters is not affected by dissolution, but only by diffusion, and if deposition is negligible, Eq. (40) reduces to Eq. (30):

$$\left.\frac{\partial C}{\partial t}\right|_x = D_s \frac{\partial^2 C}{\partial x^2}$$

This is a simple diffusion equation and integration for the boundary conditions

$x = 0 \qquad C = 0$

$x \to \infty \qquad C \to C_i$

yields (Carslaw and Jaeger, 1959)

$$C = C_i \operatorname{erf}\left(\frac{x}{2\sqrt{D_s t}}\right) \tag{42}$$

where $\operatorname{erf} \equiv$ error function $= (2/\sqrt{\pi}) \int_0^v \exp(-\epsilon^2)\, d\epsilon$. The upward flux at the sediment-water interface is

$$-J_{x=0} = \Phi_0 D_s \frac{\partial C}{\partial x}\bigg|_{x=0} = \Phi_0 C_i \left(\frac{D_s}{\pi t}\right)^{\frac{1}{2}} \tag{43}$$

The total mass M per unit area added to the surface in a given time is given by

$$\bar{M} = \int_0^t -J_{x=0}\, dt = \Phi_0 C_i \left(\frac{D_s}{\pi}\right)^{\frac{1}{2}} \int_0^t t^{-\frac{1}{2}}\, dt \tag{44}$$

$$= 2\Phi_0 C_i \left(\frac{D_s t}{\pi}\right)^{\frac{1}{2}} \tag{45}$$

If

$\Phi_0 = 0.90$

$C_i = 5$ ppm Fe^{++} or 5×10^{-6} g Fe^{++}/cm^3

$D_s = 10^{-5}$ cm^2/sec

$t = 100$ years $= 3.1 \times 10^9$ sec

then

$\bar{M} = 9 \times 10^{-4}$ g Fe/cm^2

Total solids deposited in 100 years based on measured rates of deposition range from about 8 g/cm^2 for shallow-water terrigenous muds to 8×10^{-3} g/cm^2 for deep-sea clay. From these values and the value of $\bar{M}$ above, it is apparent that for the model of pure diffusion without dissolution (Eq. 45) the assumption that monomineralic layers can be formed undiluted by deposited sediment generally is not valid. If there were absolutely no deposition, over 10 million years would be required to form a pure goethite layer (with $\Phi = 0.90$) only 1 cm thick. Therefore, in order to build up an appreciable thickness with a high concentration of iron (or manganese), equations such as Eq. (40) must be used which represent the type of model where Fe^{++} and Mn^{++} are constantly supplied to the sediment pore waters by dissolution from scattered Fe- and Mn-containing particles.

Formation of pyrite layers by diffusion during early diagenesis has been discussed by Berner (1969). The suggested mechanism is shown in Fig. 6-8. An organic-rich layer undergoing bacterial decay overlies organic-poor sediment and as a result diffusion of reduced bacterial metabolites into the underlying sediment takes place. If the sediment is moderately high in reactive iron, H$_2$S formed in the organic layer by sulfate-reducing bacteria, in

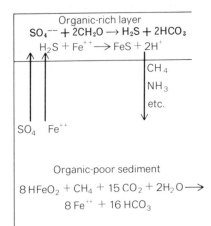

Fig. 6-8 Diffusion model for iron sulfide layer formation. Arrows represent diffusion directions. (*After Berner, 1969.*)

contrast to other metabolites, is initially trapped because it is immediately precipitated as FeS before it can diffuse out of the layer. Meanwhile iron reduction in the underlying sediment by the downward-diffusing reduced decay products (which is probably mediated by bacteria) produces a high dissolved Fe^{++} content. This Fe^{++} diffuses upward to the organic layer where it is precipitated by H_2S. Simultaneously SO_4^{--} diffuses upward to the organic layer to replace that which has been reduced by bacteria to H_2S. The continued upward diffusion of SO_4^{--} and Fe^{++} and reduction of SO_4^{--} to H_2S, leads to a buildup of precipitated FeS at the boundary between the organic layer and underlying sediment. The FeS is later transformed to pyrite.

Pyrite layers can also form by the later sulfidation of hydrous ferric oxide layers originally formed by oxidation and precipitation of Fe^{++} at the sediment-water interface as discussed earlier.

CONCRETION FORMATION

The rate of formation of spherical concretions from flowing and nonflowing ground water has been calculated (Berner, 1968b) by analogy with the growth of spherical crystals suspended or falling through a supersaturated solution (Frank, 1950; Nielsen, 1961).   The growth of hypothetical spherical crystals by diffusion is given by Eq. (88) of Chapter 2:

$$\frac{dr}{dt} = \frac{vD_s(C_\infty - C_s)}{r} \tag{46}$$

The same equation can be applied to spherical concretions which grow by diffusion from nonflowing sediment pore water.   Integrating and solving for t

$$t = \frac{r^2}{2vD_s(C_\infty - C_s)} \tag{47}$$

Thus, if D_s and the supersaturation $(C_\infty - C_s)$ are known, the time to grow a concretion of radius r can be calculated.

If supersaturated water is flowing past the growing concretion, dissolved material is added not only by diffusion but by mass flow.   The velocity of flow u in sediments is very low and, therefore, presumably laminar.   The case of growth of a spherical body by diffusion and uniform laminar flow is treated by Nielson (1961), and an integrated equation appropriate to concretions (Berner, 1968b) is

$$t = \frac{(r - D_s/0.715u)(1 + ru/D_s)^{0.715} + D_s/0.715u}{1.715uv(C_\infty - C_s)} \tag{48}$$

An example of the use of Eqs. (47) and (48) is shown in Fig. 6-9 for the growth of calcite concretions.   Note the relative unimportance of low flow rates on the time for growth.

One causative factor in the formation of pyrite concretions has been postulated by the author (Berner, 1969) to be the same as that mentioned earlier for pyrite layers, which is the diffusion of Fe^{++} and SO_4^{--} to organic-rich regions (decaying organisms, etc.) where SO_4^{--} is reduced to H_2S. (This may also explain the common occurrence of pyritized fossils.)   The causative factor in the formation of most calcite concretions is probably heterogeneous nucleation by shell fragments, etc., although alkaline putrefaction of decaying organisms rich in basic proteins, such as fish, may be an

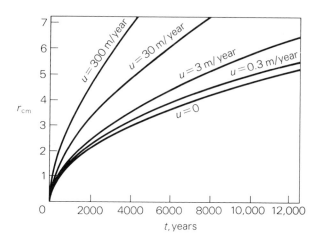

Fig. 6-9 Growth curves for spherical calcite concretions, $v = 35$ cm^3/mole, $(C_\infty - C_s) = 10^{-7}$ mole/cm^3 = 10 ppm dissolved $CaCO_3$, $D_s = 10^{-5}$ cm^2/sec. The symbol u refers to ground water flow velocity; r refers to concretion radius. (*After Berner, 1968b.*)

important precipitation mechanism during earliest diagenesis (Weeks, 1957). The first-formed precipitate during alkaline putrefaction may not be $CaCO_3$ but, instead, calcium soaps or salts of fatty acids. Experimental and field evidence for the formation of natural calcium soaps, called adipocere, by the decomposition of fish has been demonstrated (Berner, 1968a). The general process of precipitation is

$$NH_3 + RCOOH \rightarrow NH_4^+ + RCOO^-$$
$$\text{fatty acid}$$

where R = hydrocarbon chain, followed by

$$Ca^{++} + 2RCOO^- \rightarrow Ca(RCOO)_2$$

In the presence of excess free fatty acid, calcium soaps of myristate, palmitate, and stearate are more stable than $CaCO_3$. In other words for the reaction

$$2RCOOH + CaCO_3 \rightarrow Ca(RCOO)_2 + CO_2 + H_2O$$

ΔF is much less than zero under sedimentary conditions of temperature and CO_2 concentration. Once the adipocere concretion is formed and all free fatty acid is used up, the Ca soap becomes thermodynamically unstable relative to $CaCO_3$ + hydrocarbons. Given sufficient time, the soap may then break down and result in a $CaCO_3$ concretion.

DONNAN EQUILIBRIUM

The electrical double layers on clay particles are primarily a result of atomic substitutions within the crystal lattice which give rise to a net negative charge,

Table 6-2   Thickness of the Gouy layer as a function of electrolyte concentration for constant surface charge (type I electrical double layer)
Charge density equals 11.7 microcoulombs/cm² (a typical value for montmorillonite). Thickness is equal to the distance where the electrical potential Ψ falls to a value $1/e$ times the surface value.   [After Van Olphen (1963, p. 256)]

	Thickness	
Normality	1–1 valent	2–2 valent
$10^{-5} N$	1000 Å	500 Å
$10^{-3} N$	100 Å	50 Å
$10^{-1} N$	10 Å	5 Å

although a much smaller positive charge on the edges of clay platelets may result from broken bonds and the absorption of potential determining ions. Thus, clay minerals in the main have negative double layers of type I according to the classification in Chapter 2. The thickness of the Gouy layer on a clay is variable and depends upon the valence and concentration of cations in the external solution. Values of thickness, defined roughly as the distance from the surface to the point where the electrical potential falls to $1/e$ times the potential at the surface, according to Gouy theory for a typical clay, are shown in Table 6-2. At low electrolyte (cation) concentrations characteristic of fresh waters the thickness is large and on the order of the same distance as the dimensions of the clay particles. At higher concentrations typical of sea water the thickness is on the order of only a few angstroms.

If during compaction clay particles are packed so closely together that the Gouy layers interpenetrate, then the interstitial water of the sediment may show Donnan phenomena. The degree of compaction necessary for this to occur depends on the thickness of the Gouy layer which in turn depends on the salinity of the pore water (Table 6-2). In dilute (fresh) pore water Donnan phenomena may occur in surficial loosely compacted sediments, as has been demonstrated for soils (Marshall, 1964). On the other hand, at sea-water salinity a high degree of compaction is necessary so that the thin double layers may interpenetrate. This should occur only after burial has produced an average porosity of about 0.3 (lower values for pure illite and kaolinite; higher values for montmorillonite). In other words a fine-grained marine sediment needs to be largely converted to shale before Donnan effects predominate.

Once interstitial water begins to consist mainly of interpenetrating Gouy layers the Donnan equilibrium, Eq. (109) of Chapter 2, is applicable:

$$\bar{a}^+ \bar{a}^- = a_{ex}^+ a_{ex}^- \tag{49}$$

where $\bar{a}$ = space-averaged activity in the interstitial (Gouy) water

a_{ex} = activity in the overlying water or in the interstitial waters of adjoining layers, such as sands, which exhibit no Donnan effects

The anion most useful for studying the Donnan equilibrium is Cl_{aq}^-, which does not enter into chemical diagenetic reactions except in the most saline sediments. Therefore, let $a^- = a_{Cl^-}$. The most common cation in sea water is Na^+ so that for the following calculation it is assumed that $a^+ = a_{Na^+}$. The Donnan equilibrium for Na_{aq}^+ and Cl_{aq}^- is

$$\bar{a}_{Na^+} \bar{a}_{Cl^-} = a_{Na^+} a_{Cl^-} \tag{50}$$

where the subscript "ex" has been dropped for the overlying or adjoining water. Equation (50) can be rewritten in terms of molalities and activity coefficients:

$$\bar{\gamma}_{Na^+} \bar{\gamma}_{Cl^-} \bar{m}_{Na^+} \bar{m}_{Cl^-} = \gamma_{Na^+} \gamma_{Cl^-} m_{Na^+} m_{Cl^-} \tag{51}$$

Also, for simplification assume that Na^+ and Cl^- are the only ions present. Electroneutrality requires

$$m_{Na^+} = m_{Cl^-} \tag{52}$$

$$\bar{m}_{Na^+} = \bar{m}_{Cl^-} + \bar{m}_{clay^-} \tag{53}$$

where $\bar{m}_{clay^-}$ = concentration of negatively charged clay sites per kilogram of pore water.

Combining Eqs. (51), (52), and (53)

$$\bar{m}_{Cl^-}^2 + \bar{m}_{clay^-}\bar{m}_{Cl^-} - m_{Cl^-}^2\left[\frac{\gamma_{Na^+}\,\gamma_{Cl^-}}{\bar{\gamma}_{Na^+}\,\bar{\gamma}_{Cl^-}}\right] = 0 \tag{54}$$

Solving for $\bar{m}_{Cl^-}$

$$\bar{m}_{Cl^-} = \frac{-\bar{m}_{clay^-}}{2} + \frac{1}{2}\sqrt{\bar{m}_{clay^-}^2 + 4m_{Cl^-}^2\left[\frac{\gamma_{Na^+}\,\gamma_{Cl^-}}{\bar{\gamma}_{Na^+}\,\bar{\gamma}_{Cl^-}}\right]} \tag{55}$$

From this equation, the degree of anion exclusion is a function of the relative magnitude of $(\bar{m}_{clay^-}^2)$ and $(4m_{Cl^-}^2-[\gamma_{Na^+}\gamma_{Cl^-}/\bar{\gamma}_{Na^+}\bar{\gamma}_{Cl^-}])$. Assuming that the activity coefficient quotient is close to one, as

$$4m_{Cl^-}^2/\bar{m}_{clay}^2 \to 0$$

$$\bar{m}_{Cl^-} \to 0$$

This situation, anion exclusion, is favored by dilute external water (low m_{Cl^-}) and/or high $\bar{m}_{clay^-}$ brought about by a high clay content, high compaction, or high charge on the clay. Conversely, as

$$\bar{m}_{clay^-}^2/4m_{Cl^-}^2 \to 0$$

$$\bar{m}_{Cl^-} \to m_{Cl^-}$$

and little anion exclusion results.*   This situation is favored by saline external water (high m_{Cl^-}) and/or low $\bar{m}_{clay^-}$ due to a low clay content, little compaction, or low charge on the clay.

Adjacent sediment layers of different clay content or mineralogy, when sufficiently compacted so that Donnan phenomena become important, may exhibit different degrees of anion exclusion.   If so, the concentration of ions in the pore waters of each layer may be different *at equilibrium*.   In this case concentration gradients would be a stable phenomenon and not subject to erasure by diffusion.   Thus, care should be exercised in inferring diffusion in fresh water muds or soils (where Donnan phenomena may exist) from the presence of gradients in the concentration of dissolved species.

* However, $\bar{m}_{Cl^-}$ can never equal m_{Cl^-}, because the difference $(m_{Cl^-} - \bar{m}_{Cl^-})$ increases with increasing m_{Cl^-} (Van Olphen, 1963).

Modern marine sediments because of their high porosity (>0.6) and high ionic strength should exhibit little Donnan phenomena. This is because at $I = 0.7$, the volume of water included within the Gouy layer is negligible compared to the total interstitial water. However, compaction to shalelike porosities should lead to anion exclusion. This is proven by experimental studies where clay-electrolyte gels have been compacted at high pressures (McKelvey and Milne, 1962; Hanshaw, 1962). Water forced through the compacted clay emerges with a dissolved salt content lower than that at the start. This phenomena, known as salt filtering, is brought about by the partial exclusion of anions from the compacted clay. If anions are excluded, cations must also be excluded in order to maintain electroneutrality, and as a result, dissolved salt is filtered from the water by the compacted clay which acts as a semipermeable membrane.

As a result of salt filtration, dissolved salts build up in a water at the point where it enters a compacted clay exhibiting Donnan behavior. If the laboratory results can be extrapolated to nature, it is possible that dissolved salts also build up in underlying sandstones as a natural ground water passes upward through semipermeable shale beds. Highly saline brines are typical of deep sandstone formation waters of sedimentary basins and they may arise as a consequence of salt-filtering (Engelhardt and Gaida, 1963; Bredehoeft et al., 1963; Graf et al., 1965; Hitchon and Friedman, 1969). A major problem presented by this hypothesis is whether natural hydrostatic heads are sufficiently high to enable water to be forced upward through shale, which if compacted enough to bring about salt filtering must also be highly impermeable and subject to reverse pressure gradients arising from osmosis. The effectiveness of salt filtering as a process for the formation of subsurface brines has been questioned recently by Manheim and Horn (1968) on this basis. Nevertheless, highly saline deep waters are very common and, if derivation by salt filtering is rejected as a hypothesis, other sources of dissolved ions are required. The alternative sources are less likely than salt filtering. An original high salinity is improbable because of the negligible overall volumetric abundance of evaporite brines, as compared to sea water. Derivation of salts by the dissolution of buried evaporite beds is more likely, but the work of Rittenhouse (1967) indicates that halite dissolution should lead to a low Br^-/total-dissolved-salt ratio, and this is not exhibited by most subsurface brines. Whether deeply buried brines may arise from natural salt filtration by shales remains an open question.

REFERENCES

Berner, R. A., 1968a, Calcium carbonate concretions formed by the decomposition of organic matter, *Science*, v. 159, pp. 195–197.
———, 1968b, Rate of concretion growth, *Geochim. et Cosmochim. Acta*, v. 32, pp. 477–483.

————, 1969, Migration of iron and sulfur within anaerobic sediments during early diagenesis, *Am. Jour. Sci.*, v. 267, pp. 19–42.

Bredehoeft, J. D., Blyth, C. R., White, W. A., and Maxey, G. B., 1963, Possible mechanism for concentration of brines in subsurface formations, *Am. Assoc. Petroleum Geologists Bull.*, v. 47, pp. 257–269.

————, and Hanshaw, B. B., 1968, On the maintenance of anomalous fluid pressures: I. Thick sedimentary sequences, *Geol. Soc. Amer. Bull.*, v. 79, pp. 1097–1106.

Carslaw, H. S., and Jaeger, J. C., 1959, *Conduction of heat in solids*, Clarendon Press, Oxford, 510 p.

Crank, J., 1956, *The mathematics of diffusion*, Clarendon Press, Oxford, 347 p.

Dickinson, G., 1953, Geological aspects of abnormal reservoir pressures in Gulf Coast Louisiana, *Am. Assoc. Petroleum Geologists Bull.*, v. 37, pp. 410–432.

Duursma, E. K., and Hoede, C., 1967, Theoretical, experimental, and field studies concerning molecular diffusion of radioisotopes in sediments and suspended solid particles of the sea, Part A. Theories and mathematical calculation, *Netherlands Jour. Sed. Research*, v. 3, pp. 423–457.

Emery, K. O., 1960, *The sea off southern California*, Wiley, New York, 366 p.

————, and Rittenberg, S. C., 1952, Early diagenesis of California basin sediments in relation to origin of oil, *Am. Assoc. Petroleum Geologists Bull.*, v. 36, pp. 735–806.

Engelhardt, W. V., 1960, *Der Porenraum der Sedimente*, Springer-Verlag, Berlin, 207 p.

————, and Gaida, K. H., 1963, Concentration changes of pore solutions during the compaction of clay sediments, *Jour. Sed. Petrology*, v. 33, pp. 919–930.

Frank, F. C., 1950, Radially symmetric phase growth controlled by diffusion, *Proc. Roy. Soc. London*, Series A, v. 201, pp. 586–599.

Friedman, G. M., 1964, Early diagenesis and lithification in carbonate sediments, *Jour. Sed. Petrology*, v. 34, pp. 777–813.

Gast, R. G., 1963, Relative effects of tortuosity, electrostatic attraction and increased viscosity of water on self-diffusion rates of cations in bentonite-water systems, *Proceedings of the 1963 International Clay Conf.*, Pergamon, New York, pp. 251–259.

Goodknight, R. C., Klikoff, W. A., and Fatt, I., 1960, Non steady-state fluid flow and diffusion in porous media containing dead-end pore volume, *Jour. Phys. Chem.*, v. 64, pp. 1162–1168.

Graf, D. L., Friedman, I., and Meents, W. F., 1965, The origin of saline formation waters, II. Isotopic fractionation by shale micropore systems, *Illinois State Geol. Surv. Circular* 393.

Hanshaw, B. B., 1962, Membrane properties of compacted clays, unpublished Ph.D. thesis, Harvard University, 113 p.

Heald, M. T., 1955, Stylolites in sandstones, *Jour. Geology*, v. 63, pp. 101–114.

Hedberg, H. D., 1936, Gravitational compaction of clays and shales, *Am. Jour. Sci.*, v. 31, pp. 241–287.

Hitchon, B., and Friedman, I., 1969, Geochemistry and origin of formation waters in the western Canada sedimentary basin, I. Stable isotopes of hydrogen and oxygen, *Geochim. et Cosmochim. Acta*, v. 33, pp. 1321–1350.

Lai, T. M., and Mortland, M. M., 1962, Self-diffusion of exchangeable cations in bentonite, *Clays and Clay Minerals*, 9th Conf., Pergamon, New York, pp. 229–247.

Mackenzie, F. T., Ginsburg, R. N., Land, L. S., and Bricker, O. P., 1969, Carbonate cements, *Bermuda Biological Station for Research Special Publication No. 3*, 325 p.

Manheim, F. T., and Bischoff, J. L., 1969, Geochemistry of pore waters from Shell Oil Co. drill holes on the continental slope of the northern Gulf of Mexico, *Chemical Geology*, v. 4, pp. 63–82.

————, and Horn, M. K., 1968, Composition of deeper subsurface waters along the Atlantic continental margin, *Southeastern Geology*, v. 9, pp. 215–236.

Marshall, C. E., 1964, *The physical chemistry and mineralogy of soils*, Wiley, New York, 388 p.

McKelvey, J. G., and Milne, I. H., 1962, The flow of salt solutions through compacted clay, *Clays and Clay Minerals*, 9th Conf., Pergamon, New York, pp. 248–259.

Meade, R. H., 1966, Factors influencing the early stages of the compaction of clays and sands—review, *Jour. Sed. Petrology*, v. 36, pp. 1085–1101.

Nielsen, A. E., 1961, Diffusion controlled growth of a moving sphere. The kinetics of crystal growth in potassium perchlorate precipitation, *Jour. Phys. Chem.*, v. 65, pp. 46–49.

Rittenhouse, G., 1967, Bromine in oil-field waters and its use in determining possibilities of origin of these waters, *Am. Assoc. Petroleum Geologists Bull.*, v. 51, pp. 2430–2440.

Scholl, D. W., and Johnson, W. L., 1967, Effects of diffusion on interstitial water chemistry in deltaic areas, *Seventh International Sedimentological Congress*.

Siever, R., 1959, Petrology and geochemistry of silica cementation in some Pennsylvania sandstones, in *Silica in Sediments*, *Soc. of Economic Paleontologists and Mineralogists Spec. Publication No. 7*, p. 55–79.

Sippel, R. F., 1968, Sandstone petrology, evidence from luminescence petrography, *Jour. Sed. Petrology*, v. 38, pp. 530–554.

Van Olphen, H., 1963, *Clay colloid chemistry*, Interscience-Wiley, 310 p.

Weeks, L. G., 1957, Origin of carbonate concretions in shales of the Magdelena Valley, Colombia, *Geol. Soc. Amer. Bull.*, v. 68, pp. 95–102.

Weller, J. M., 1959, Compaction of sediments, *Am. Assoc. Petroleum Geologists Bull.*, v. 43, pp. 273–310.

Weyl, P. K., 1959, Pressure solution and the force of crystallization—a phenomenological theory, *Jour. Geophys. Research*, v. 64, pp. 2001–2025.

Wollast, R., 1969, Kinetic aspects of the nucleation and growth of calcite from aqueous solution, in *Carbonate Cements*, Mackenzie, F. T., Ginsburg, R. N., Land, L. S., and Bricker, O. P. (eds.), *Bermuda Biological Station for Research Special Publication No. 3*, pp. 195–207.

7
Diagenetic Redox Reactions in the System C-N-S-H-O

Besides iron and manganese the common elements in sediments which exhibit multiple valence states, and thus can undergo oxidation and reduction, are carbon, nitrogen, sulfur, hydrogen, and oxygen. The purpose of this chapter is to discuss some of the more important redox reactions which take place during the early diagenesis of sediments between compounds containing these elements. Emphasis is placed on the formation by bacteria of the common dissolved species found in pore waters of recent sediments and the effect such bacterial metabolites have upon the Eh and pH of the sediments.

Eh AND REDOX EQUILIBRIUM

The elements C-N-S-H-O are closely linked with biological processes, and as a result, nonequilibrium distributions in natural waters of compounds containing these elements can occur. This is because the most important biological process, photosynthesis, utilizes light energy to produce thermo-dynamically unstable substances, i.e., organic matter. During the decomposition of organic matter, equilibrium is approached but its attainment is often dependent upon the presence of enzyme systems in bacteria and other

114

organisms which can surmount kinetic activation-energy barriers. Thus, the steady-state distribution in natural waters of substances which contain carbon, nitrogen, etc., is dependent upon the relative rates of displacement away from and catalysis toward equilibrium. The purpose of the present section is to calculate the situation expected for thermodynamic equilibrium so that the degree of approach to equilibrium of natural waters can be evaluated.

EQUILIBRIUM CALCULATIONS

There are an enormous number of known chemical compounds in the system C–N–S–H–O. Included are most organic substances and many simple inorganic species. To represent relative stabilities in this system would require a compendium longer than this book. However, for the situation of homogeneous equilibrium in natural waters, the problem can be greatly simplified. This is because at sedimentary values of temperature, Eh, and pH, and at thermodynamic equilibrium, only a small number of common dissolved species make up almost all of the total carbon, nitrogen, sulfur, hydrogen, and oxygen in aqueous solution. They are $H_{2\,aq}$, $O_{2\,aq}$, $CH_{4\,aq}$, CO_3^{--}, HCO_3^-, H_2CO_3, $NH_{3\,aq}$, NH_4^+, $N_{2\,aq}$, NO_3^-, H_2S_{aq}, HS^-, and SO_4^{--} (Thorstenson, 1970). All other species, such as $S_2O_3^{--}$, are present at equilibrium at much lower concentrations. The fields where each carbon, nitrogen, and sulfur species predominates over others containing the same element are plotted as a function of Eh and pH in Fig. 7-1. The diagram is constructed from the thermodynamic data of Appendix I for $P_{N_2} = 0.8$ atm (the value in air), $T = 25°C$, and $P_{total} = 1$ atm. Each boundary line represents the locus of points where the *activities* of two species of a given element are equal. Also included are four curves (dashed lines) representing plots of the half-cell reactions:

$$2H_2O \leftrightharpoons O_{2\,gas} + 4H_{aq}^+ + 4e$$

$$Eh = 1.23 - 0.0592\ pH + 0.0148\ \log P_{O_2} \tag{1}$$

$$H_{2\,aq} \leftrightharpoons 2H_{aq}^+ + 2e$$

$$Eh = -0.0592\ pH - 0.0296\ \log P_{H_2} \tag{2}$$

for values of P_{O_2} and P_{H_2} of 10^{-4} atm and 1 atm. The latter value represents the limit for the stability of water at $P_{total} = 1$ atm (Garrels and Christ, 1965). At higher values of P_{O_2} and P_{H_2} (areas indicated on the diagram) water is thermodynamically unstable relative to hydrogen or oxygen gas. Other boundaries are derived as shown by the following examples.

1. *Normal* Eh-pH *boundaries*:

$$CH_{4\,aq} + 3H_2O \leftrightharpoons CO_{3\,aq}^- + 10H_{aq}^+ + 8e$$

$$Eh = E° + \frac{RT}{8\mathscr{F}} \ln \frac{a_{CO_3^{--}}\, a_{H^+}{}^{10}}{a_{CH_4}} \tag{3}$$

If $a_{CH_4} = a_{CO_3^{--}}$ and $T = 25°C$

$$Eh = E° - 0.0740 \text{ pH} \tag{4}$$

2. Eh-pH *boundary with* $N_{2 \text{ aq}}$ *(dimer)*:

$$2NH_{3 \text{ aq}} \leftrightharpoons N_{2 \text{ aq}} + 6H_{aq}^+ + 6e$$

$$Eh = E° + \frac{RT}{6\mathscr{F}} \ln \frac{a_{N_2} a_{H^+}^6}{a_{NH_3}^2} \tag{5}$$

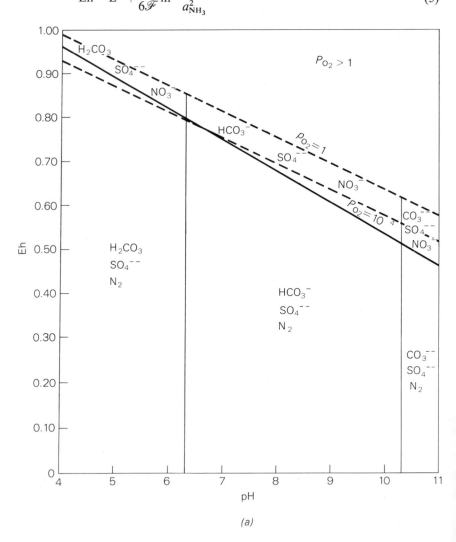

(a)

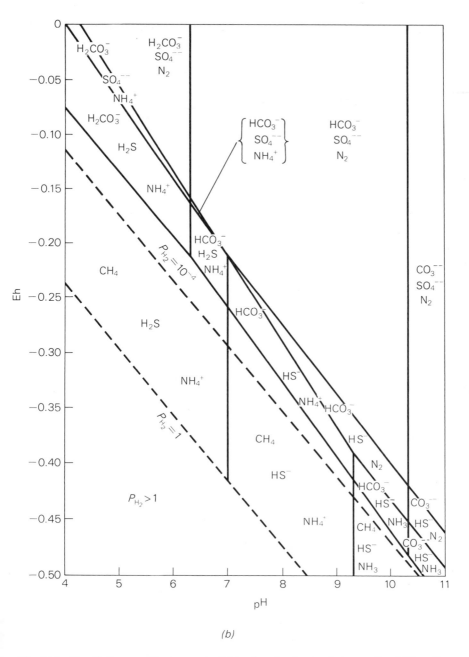

Fig. 7-1 Eh-pH diagrams illustrating the fields of predominance, in terms of activities, of dissolved species in the system C–N–S–H–O. The diagrams are constructed for the extreme pH range of sediments (4 to 11), $T = 25°C$, $P_{total} = 1$ atm, and $P_{N_2} = 0.8$ atm (the value for air). Areas marked $P_{O_2} > 1$ and $P_{H_2} > 1$ refer to places where water is thermodynamically unstable relative to $O_{2\ gas}$ and $H_{2\ gas}$ respectively. (a) Eh range 0 to 1.00; (b) Eh range 0 to −0.50.

If $a_{N_2} = a_{NH_3}$ and $T = 25°C$

$$Eh = E° - 0.0592 \text{ pH} - 0.010 \log a_{N_2} \qquad (6)$$

If $P_{N_2} = 0.8$ atm, $a_{N_2} = 10^{-3.3}$

Thus

$$Eh = (E° + 0.033) - 0.0592 \text{ pH} \qquad (7)$$

3. Eh *independent boundaries*:

   Example:

   $$HCO_{3\,aq}^- \rightleftharpoons CO_{3\,aq}^{--} + H_{aq}^+$$

   $$K = \frac{a_{H^+} a_{CO_3^{--}}}{a_{HCO_3^-}} \qquad (8)$$

   If $a_{HCO_3^-} = a_{CO_3^{--}}$

   $$pH = -\log K = pK \qquad (9)$$

   From Fig. 7-1 some predictions can be made.

1. The principal species in aerobic surface waters in equilibrium with the atmosphere ($P_{O_2} = 0.2$ atm) at neutral pH should be HCO_3^-, NO_3^-, and SO_4^{--}.
2. In anaerobic waters ($P_{O_2} < 10^{-4}$), at pH values (7 to 8) most commonly encountered in marine sediments, the sequence of coexisting predominant dissolved species during reduction and drop in Eh should be $HCO_3^- + SO_4^{--} + NO_3^-$; $HCO_3^- + SO_4^{--} + N_2$; $HCO_3^- + HS^- + N_2$; $HCO_3^- + HS^- + NH_4^+$; $CH_4 + HS^- + NH_4^+$. This means that O_2 reduction should be followed by NO_3^- reduction, then SO_4^{--} reduction, then N_2 reduction, and finally HCO_3^- reduction.
3. The dissolved species to be expected in the most reduced waters at pH 7 to 8 are $CH_4 + HS^- + NH_4^+ + H_2$.

Actual occurrences in natural waters are in general agreement with these predictions although specific exceptions occur. Detailed discussion follows.

AEROBIC WATERS

Natural waters containing measurable concentrations of dissolved O_2 ($P_{O_2} > 10^{-4}$ atm) are here defined as being aerobic. These include all surface waters and almost all bottom waters. The few exceptions include bottom waters of some lakes, and bottom waters of restricted basins of the ocean, the best known example being the Black Sea. The open ocean is aerobic at all depths. The predominant dissolved species in the system C–N–S–H–O found in aerobic waters are HCO_3^-, SO_4^{--}, and N_2. Dissolved NO_3^- is minor, although at equilibrium with $O_{2\,aq}$ it should be much more abundant than N_2 (see Fig. 7-1). Apparently, biogeochemical reactions involving nitrogen result in a steady-state concentration of dissolved N_2 in aerobic water which is far higher than that predicted for thermodynamic equilibrium.

The Eh of aerobic waters, measured by using platinum electrodes, does not agree at the measured pH with that predicted by Eq. (1) for the reversible O_2–H_2O half-cell at $P_{O_2} = 0.2$ atm. This can be explained partly by the very low exchange current for this half cell at a platinum surface (Stumm, 1966). The potential response of platinum in aerated water is similar to that found in weak solutions of hydrogen peroxide (Bockris and Oldfield, 1955; Sato, 1960). Bockris and Oldfield found that in the presence of H_2O_2 at concentrations greater than 10^{-6} molar, platinum and gold electrodes obeyed the empirical equation

$$Eh = 0.84 - 0.0592\ pH$$

which is approximately the same relationship found in natural fully aerated waters (which should contain traces of H_2O_2) by Sato (1960) and others. The lack of dependency of Eh upon H_2O_2 activity above 10^{-6} m was explained by Bockris and Oldfield to be due to the adsorption of OH° radicals (or H_2O_2) over the entire surface of the electrode, resulting in surface poisoning. If true, this poisoning explains why the electrode exhibits a theoretical pH response, while at the same time it fails to measure P_{O_2}. The suggested potential controlling mechanism is

$$H_2O \leftrightharpoons OH^\circ_{ads} + H^+ + e$$

It is interesting that the platinum electrode, although presumably poisoned, gives an Eh for aerobic waters at neutral pH which predicts the correct dissolved nitrogen species N_2, whereas the assumption of $P_{O_2} = 0.2$ atm does not.

ANAEROBIC WATERS

Anaerobic waters are confined mainly to the interstices of sediments and are common there. The lack of dissolved O_2 is due to the oxidation of sedimented organic matter and reduced metabolites by aerobic microorganisms. Due to a high metabolic rate, aerobic bacteria can use up oxygen very rapidly. As a result, aerobic waters are often underlain by anaerobic sediments.

The aerobic bacteria live at the top surface of the sediment and by their rapid utilization prevent the diffusion of oxygen from above into the sediment. This continues as long as decomposable organic matter is supplied to the sediment faster than it can be destroyed by aerobic oxidation. The environment which best favors this is one of quiet water where low density organic detritus can settle out and not be continually resuspended and removed by currents or oxidized by aerobic bacteria. Therefore, anaerobic waters are more closely associated with silts and clays than with sands.

Once dissolved O_2 is removed, any further oxidation of organic matter by bacteria must occur by the utilization of other oxidized dissolved species. These include NO_3^-, SO_4^{--}, CO_2, and oxidized organic species. The succession of bacterial processes observed by Richards (1965) and others in marine waters (pH 7 to 8) is nitrate reduction, followed by SO_4^{--} reduction plus ammonia formation, and finally, methane formation. This is in agreement with the succession predicted in Fig. 7-1 for continually decreasing Eh at neutral pH. The direct reduction of N_2 to NH_4^+ by bacteria, however, is not documented (Thimann, 1963). Most ammonia comes instead from the decomposition of organic nitrogen compounds.

The chief C–N–S–H–O dissolved species found in most sulfide-rich sediments (pH 7 to 8) are HCO_3^-, NH_4^+, HS^-, H_2S, CH_4, N_2, and SO_4^{--}. Measurements of Eh, however, do not reflect the Eh expected for any of the half-cell couples using combinations of these species. This is to be expected since SO_4^{--}, N_2, CH_4, and HCO_3^- are not electroactive (rapidly reactive at an electrode surface). The measured Eh in sulfide-rich sediments, however, is meaningful and is controlled by other less abundant but more electroactive species. Berner (1963) has found that the measured Eh of many sulfidic sediments is the same as that predicted for the couple

$$HS^- \leftrightharpoons S° + H^+ + 2e$$

The feasibility of this half cell as an electrode controlling mechanism has been shown by high exchange currents both in laboratory solutions (Allen and Hickling, 1957) and in natural sediments (Doyle, 1967). The electroactive species are HS^- and a series of polysulfide ions S_2^{--}, S_3^{--}, etc., which form by the reaction of $S°$ with HS^-. Reversibility has also been demonstrated by Voge (1939) who found rapid exchange of radioactive S^{35} between dissolved sulfide and sulfur.

Although Eh as measured with a platinum electrode is not affected by redox couples involving the principal dissolved C–N–S–H–O species, it is still possible to check for internal equilibrium by *calculating* Eh from measured concentrations and comparing the resulting values for each carbon, nitrogen, and sulfur couple. The Eh for all pairs should be the same if there is complete internal equilibrium. Calculation of Eh for each of the pairs SO_4^{--}–HS^-, HCO_3^-–CH_4, and N_2–NH_4^+ from published concentrations for several different areas and from new measurements of Bermuda sediments has been done by Thorstenson (1970). The results are shown in Fig. 7-2. The calculated Eh values for SO_4^{--}–HS^- are in much better agreement with those for N_2–NH_4^+, as can be seen by a clustering around the 1:1 correlation line, than Eh values calculated for HCO_3^-–CH_4. An idea of the degree of divergence from equilibrium is that a 0.002 to 0.010 volt error shown for Eh_S

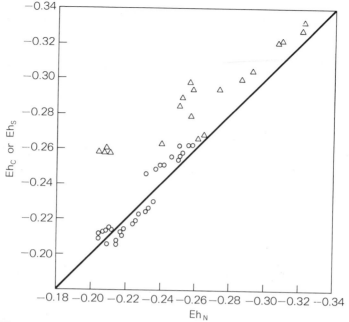

Fig. 7-2 Values of Eh calculated from measured concentrations of: $N_{2\,aq}$ and $NH_{4\,aq}^{+}(Eh_N)$; $SO_{4\,aq}^{--}$ and $HS_{aq}^{-}(Eh_S)$; and $HCO_{3\,aq}^{-}$ and $CH_{4\,aq}(Eh_C)$. (*Modified from Thorstenson, 1970.*) The straight line is the locus of all points corresponding to complete internal equilibrium. ○Eh_S versus Eh_N; △Eh_C versus Eh_N.

corresponds to a SO_4^{--}/HS^- ratio which is 2 to 25 times too high or too low for equilibrium with the measured concentrations of N_2 and NH_4^+. Similarly the divergence for Eh_C corresponds to a HCO_3^-/CH_4 ratio which is from 3 to 10,000,000 times too low. Thus, although the data indicate an approach to internal equilibrium, there is not complete equilibration, especially in the case of HCO_3^-–CH_4. This might be expected since none of the carbon, nitrogen, or sulfur species readily exchange electrons with one another inorganically at room temperature. Specific catalysts are necessary for the attainment of equilibrium and apparently bacterial enzyme systems only partially complete the process. Actually it is somewhat amazing, considering the many different sources of each dissolved species, that equilibrium is approached as closely as it is.

Eh measurements made with a platinum electrode of anaerobic sediments *very low in dissolved sulfide* usually represent irreversible or mixed potentials (Stumm, 1966), which cannot be interpreted theoretically. This is due to a lack of abundant electroactive species which form reversible redox pairs. However, in some iron-rich pore waters the weakly reversible couple Fe^{++}–$Fe(OH)_3$ colloid may give meaningful results (Doyle, 1967).

EARLY DIAGENESIS OF CARBON

The purpose of this section is not to discuss the organic geochemistry of sediments, but rather to focus upon processes of formation, during early diagenesis, of the simple molecules, CO_2 and CH_4. For a detailed discussion of organic geochemistry the reader is referred to other books such as that by Eglinton and Murphy (1969).

FORMATION OF CO2

Oxidized carbon is added to natural waters by the biological decomposition of organic matter. It is normally added as $CO_{2\,aq}$ or $H_2CO_{3\,aq}$, but as a result of bacterial sulfate reduction it can also be added as $HCO_{3\,aq}^-$. The biological reactions of greatest quantitative importance are

1. Oxidation by aerobic organisms using $O_{2\,aq}$
2. Fermentation by bacteria and other microorganisms
3. Reduction of $SO_{4\,aq}^{--}$ by bacteria

The oxygenation of organic carbon by biological respiration is an important process for the formation of CO_2 in aerobic waters such as the oceans. The overall reaction, using carbohydrate as an example, is

$$C_6H_{12}O_6 + 6O_2 \;\rightarrow\; 6CO_2 + 6H_2O$$

In natural water both aerobic bacteria and higher organisms are responsible for reactions of this type (ZoBell, 1946). Because CO_2 forms a weak acid, H_2CO_3, the effect of aerobic decomposition is to cause a lowering of pH. In shallow waters where light penetration is sufficient to enable photosynthesis by phytoplankton, excess CO_2 produced by respiration is used up by the phytoplankton. Photosynthesis is geochemically the opposite of respiration and results in the uptake of CO_2 and production of organic matter and O_2. In deeper waters below the photic zone no photosynthesis can take piace and thus there is a net production of CO_2. As a result the pH of deeper waters is lower than shallow waters both in lakes and in the ocean.

In anaerobic sediments CO_2 cannot be formed by reaction of organic carbon with molecular O_2. Instead microbiological fermentation reactions result in the oxidation of organic carbon by the oxygen contained in organic matter. Fermentation reactions are of many types but it can be stated that all result in a lower free-energy change than that resulting from oxidation by O_2. For example, for the overall reactions

$$C_6H_{12}O_6 + 6O_{2\,gas} \;\rightarrow\; 6CO_{2\,gas} + 6H_2O_{liq}$$

$$C_6H_{12}O_6 \;\rightarrow\; 3CO_{2\,gas} + 3CH_{4\,gas}$$

the values of $\Delta F°$ for each reaction are, respectively, -688 and -101 kcal/mole of sugar.   The greater energy yield for aerobic oxidation results in more efficient metabolism, and thus faster decomposition of organic matter than that which occurs during fermentation.   As a result organic matter is often almost completely destroyed in waters where constant aeration takes place.   By contrast in anaerobic sediments some organic carbon escapes bacterial destruction and is eventually transformed into coal, oil, or kerogen.

The effect on pH during anaerobic fermentation depends upon the amounts of CO_2 (and organic acids) formed relative to the amounts of nitrogenous bases.   The fact that the top portions of anaerobic sediments in general have a lower pH than overlying aerobic waters indicates that fermentation reactions (plus sulfate reduction) collectively result in net acid production.   Very low values of pH (<5) can result from fermentation in swamps and bogs where the organic compounds are low in nitrogen (e.g., lignin and carbohydrates) and where no $CaCO_3$ or other basic minerals are present to buffer the pH and neutralize excess carbonic and other acids.

The bacterial reduction of sulfate results in the formation of bicarbonate ion.   This is because a divalent ion, SO_4^{--}, is converted to a monovalent ion HS^- or a neutral dissolved species, H_2S, and the excess negative charge must be maintained.   The bacteria that reduce sulfate (genus *Desulfovibrio*) utilize organic carbon as a reducing agent and the resulting CO_2 can supply the needed excess negative charge in the form of HCO_3^- ion.   Generalized overall reactions, using carbohydrate, are

$$2CH_2O + SO_{4\,aq}^{--} \rightarrow 2HCO_{3\,aq}^- + H_2S_{aq}$$

$$2CH_2O + SO_{4\,aq}^{--} \rightarrow HCO_{3\,aq}^- + HS_{aq}^- + CO_{2\,aq} + H_2O_{aq}$$

High HCO_3^- ion concentrations in sediment pore waters as a result of sulfate reduction are illustrated in Fig. 7-3.   The addition of HCO_3^- to anaerobic pore water as a result of bacterial sulfate reduction may bring about the precipitation of dissolved Ca^{++} as $CaCO_3$.   This is suggested by the studies

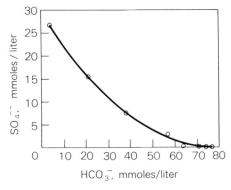

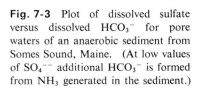

Fig. 7-3 Plot of dissolved sulfate versus dissolved HCO_3^- for pore waters of an anaerobic sediment from Somes Sound, Maine. (At low values of SO_4^{--} additional HCO_3^- is formed from NH_3 generated in the sediment.)

of Feely and Kulp (1957), Berner (1966), and Presley et al. (1968).   Maintenance of supersaturation without precipitation, however, may occur due to the
inhibiting effect upon crystallization of organic matter dissolved in the pore
water (see Chapter 4).   (Recent work by the author confirms the presence of
high degrees of supersaturation.)   The change in pH as a result of bacterial
sulfate reduction depends upon the nature of the organic source.   However,
there is some suggestion that the overall reaction in marine sediments causes
a slight increase in pH (Emery and Rittenberg, 1952; Kaplan et al., 1963),
although more recent work suggests the opposite (I. R. Kaplan, personal
communication).

METHANE FORMATION

Methane forms during early diagenesis by bacterial fermentation.   The
overall reactions for methane formation are for carbohydrates

$$2(CH_2O)_n \rightarrow nCH_4 + nCO_2$$

and for fatty acids

$$C_nH_{2n}O_2 + \frac{n-2}{2}H_2O \rightarrow \frac{n+2}{4}CO_2 + \frac{3n-2}{4}CH_4$$

The stoichiometry of the latter reaction has been verified by Thayer (1931),
who found no evidence for the formation of hydrocarbons higher than methane
as a result of the fermentation of fatty acids.   This finding is important in
that it demonstrates that microorganisms are unable to decarboxylate fatty
acids to form the common straight chain hydrocarbons found in petroleum.
Fermentation of carbohydrates is believed to be the preponderant source of
methane in most sediments (McCarty, 1964).   This is demonstrated by the
well-known abundance of methane in fresh-water swamps.   In swamps the
main source of organic matter is nonmarine plants which are rich in cellulose
and other carbohydrates.   In anaerobic marine sediments methane is also
abundant.   Reeburgh (1967) has shown that in the pore waters of muds from
Chesapeake Bay, methane is produced so rapidly that it reaches supersaturation, and as a result bubbles are formed which pass from the sediment into the
overlying water.   The bubbling process apparently causes loss from solution
of other gases, such as argon and nitrogen, by entrapment in the upward
moving methane bubbles.

   Besides fermentation, methane may also form by the reduction of carbon
dioxide, a process which is not well documented in natural environments.
According to Thimann (1963), bacteria have been shown in laboratory
experiments to reduce CO_2 to CH_4 using either $H_{2\,aq}$ or a wide variety of organic
hydrogen donors as reducing agents.   The general reaction is

$$CO_2 + 4H_2A \rightarrow CH_4 + 4A + 2H_2O$$

In the case of $H_{2\,aq}$, A is omitted.   The source of CO_2 can be an earlier fermentation reaction.   For example, successive reactions for sugar fermentation can be

I. $C_6H_{12}O_6 \xrightarrow{\text{bacteria I}} 2C_2H_5OH + 2CO_2$

II. $CO_2 + 2C_2H_5OH \xrightarrow{\text{bacteria II}} CH_4 + 2CH_3COOH$

Reaction II is the reduction of CO_2 by ethanol where $4H_2A = 2C_2H_5OH$, $4A + 2H_2O = 2CH_3COOH$, and $A = \frac{1}{2}(C_2H_2O)$.

Additional work on natural sediments is required to test for the quantitative significance of CO_2 reduction.   On the basis of Eh considerations discussed above this reaction would be expected thermodynamically.

DIAGENESIS OF NITROGEN

Nitrogen is added to sediments originally in three forms: dissolved N_2, dissolved NO_3^-, and nitrogen bound up in organic compounds.   In the absence of organic matter, dissolved NO_3^- and N_2 undergo little change since they do not interact with minerals.   In the presence of organic matter, NO_3^- can be reduced by bacteria to N_2 or ammonia.   However, organic nitrogen usually dominates the chemistry of nitrogen during diagenesis because it is almost always much more abundant than original NO_3^-.

In live organisms the bulk of nitrogen is present as proteins.   Upon death the proteins rapidly undergo a series of bacterial decomposition reactions. The first is proteolysis.   This is the hydrolysis of proteins to their constituent amino acids by bacteria which utilize proteolytic enzymes.   The free amino acids can subsequently undergo many different reactions.   The rapid bacterial decomposition of proteins and amino acids is illustrated by the experimental work of Krause (1959), who found decomposition of the major portion of original total protein in planktonic organisms over a period of only a few months.   This occurred under both aerobic and anaerobic conditions.   The rapid preferential destruction of proteins in natural waters is illustrated by comparison of organic carbon/nitrogen ratios for sediments and for organisms which furnish most of the organic matter to the sediments.   Table 7-1 shows that a considerable change in the C/N ratio occurs during sedimentation. Change in C/N with depth below 10 cm after sedimentation is much less. Apparently bacterial decomposition is so rapid that considerable nitrogen is lost while dead detritus is settling through the water column or while it is sitting near the sediment-water interface.

The bacterial decomposition of amino acids leads ultimately to the formation of ammonia.   This process, known as deamination, occurs both in the presence and absence of dissolved oxygen.   In the presence of oxygen the ammonia is subsequently oxidized by aerobic bacteria to NO_2^- and NO_3^-, the

Table 7-1   Average C/N weight
ratio of organisms and sediments.
[After Emery (1960)]

Organisms or sediments	C/N
Diatoms	5.1
Modern sediments	
0–10 cm depth	8.0
0–35 cm depth	11.2
275–325 cm depth	13.0
Ancient sedimentary rocks	14.8

process being called nitrification.   The principal forms are *Nitrosomonas* which oxidizes ammonia to nitrite and *Nitrobacter* which oxidizes nitrite to nitrate.   Inorganically the reaction

$$NH_4^+ + 2O_2 \rightarrow NO_3^- + 2H^+ + H_2O$$

is exceedingly slow at room temperature.   Thus, nitrifying bacteria catalyze the reaction.   Because ammonia is rapidly oxidized in natural aerobic waters, it does not accumulate in sufficiently high concentrations to alter the pH.

Under anaerobic conditions in sediments, ammonia may accumulate and reach appreciable concentrations in the interstitial water.   This has been demonstrated by the work of Rittenberg et al. (1955), where concentrations exceeding 10^{-3} M have been found.   At such concentrations ammonia can raise pH and/or bring about the precipitation of dissolved Ca^{++}.   The major reactions are

$$NH_3 + CO_2 + H_2O \rightarrow NH_4^+ + HCO_3^-$$

$$NH_3 + HCO_3^- \rightarrow NH_4^+ + CO_3^{--}$$

$$NH_3 + RCOOH \rightarrow NH_4^+ + RCOO^-$$

where   R = hydrocarbon chain

RCOOH = fatty acid

followed by

$$2HCO_3^- + Ca^{++} \rightarrow CaCO_3 + CO_2 + H_2O$$

$$CO_3^{--} + Ca^{++} \rightarrow CaCO_3$$

$$2RCOO^- + Ca^{++} \rightarrow Ca(RCOO)_2$$

The latter reactions in each set are, respectively, saponification and precipitation of calcium soap.   In the presence of an excess of free fatty acid, Ca-soap precipitation is to be expected thermodynamically (see Chapter 6, Concretion Formation).   The effect of ammonia formation, from the anaerobic decomposition of fresh fish protein, upon the concentrations of Ca^{++} and HCO_3^- is shown in Table 7-2.   Sulfate reduction and ammonia formation constitute the two major processes whereby Ca_{aq}^{++} may be precipitated from sediment pore waters.

The formation of ammonia from organic nitrogen compounds in anaerobic sediments can be treated quantitatively by means of a simplified model. The basic assumption is that organic nitrogen is decomposed to ammonia according to first-order kinetics.   Mathematically

$$\frac{dN}{dt} = -kN \tag{10}$$

where N = concentration of organic nitrogen which can be decomposed to
      ammonia

    t = time

    k = overall bacterial decay rate constant

If the organic nitrogen is present in a nondiffusable form, substitution of Eq. (10) into the general diagenetic equation from Chapter 6 yields

$$\frac{\partial N}{\partial t} = -kN - \omega \frac{\partial N}{\partial x} \tag{11}$$

where $\omega = dx/dt$.   For the case of steady-state diagenesis

$$\frac{\partial N}{\partial t} = 0 \quad \text{and} \quad \frac{\partial N}{\partial x} = \frac{-k}{\omega} N \tag{12}$$

Table 7-2   Effect of fish decomposition upon the chemical composition of sea water. [After Berner (1968)]

	Concentration, mmoles/liter	
Ion	Original sea water	After 135 days' rotting
Ca^{++}	8.5	0.5
Mg^{++}	45.5	10.4
NH_4^+	0.0	220
HCO_3^- $+ CO_3^{--}$	1.8	120
pH	6.5 (with fish)	8.8

Solving for the boundary condition, $N \to 0$ as $x \to \infty$:

$$N = N_0 \exp\left(\frac{-k}{\omega} x\right) \tag{13}$$

where $N_0 = N$ at $x = 0$.   Thus, from Eqs. (10) and (13),

$$\frac{dN}{dt} = -k N_0 \exp\left(\frac{-k}{\omega} x\right) \tag{14}$$

Now by mass balance the rate of organic nitrogen decomposition equals the rate of ammonia formation.   Since ammonia is diffusable, the general diagenetic equation for it is

$$\frac{\partial c}{\partial t} = D_s \frac{\partial^2 c}{\partial x^2} + k N_0 \exp\left(\frac{-k}{\omega} x\right) - \omega \frac{\partial c}{\partial x} \tag{15}$$

where c = concentration of total ammonia ($NH_3 + NH_4^+$).   For steady-state diagenesis

$$D_s \frac{\partial^2 c}{\partial x^2} + k N_0 \exp\left(\frac{-k}{\omega} x\right) - \omega \frac{\partial c}{\partial x} = 0 \tag{16}$$

Solution of this equation for the boundary conditions

$$x = 0 \qquad c = 0$$

$$x \to \infty \qquad c \to c_\infty \qquad \text{(asymptotic value)}$$

yields

$$c = \frac{N_0 \omega^2}{\omega^2 + D_s k}\left[1 - \exp\left(\frac{-k}{\omega} x\right)\right] \tag{17}$$

and

$$c_\infty = \frac{N_0 \omega^2}{\omega^2 + D_s k} \tag{18}$$

Also, from Eqs. (13) and (17),

$$c = \frac{\omega^2}{\omega^2 + D_s k}(N_0 - N) \tag{19}$$

The model can be tested using the data of Fig. 7-4 for the Santa Barbara Basin, California, obtained by Rittenberg et al. (1955).   The sediment is anaerobic and lithologically very uniform from top to bottom and appears to represent a reasonably good example of steady-state diagenesis.   Unfortunately the core is not deep enough to delineate the asymptotic value c_∞ or to derive a value for N_0.   However, the model can be tested by using the series

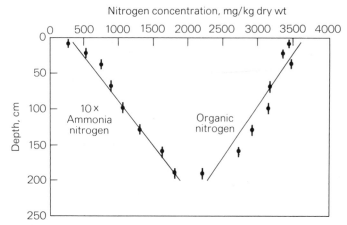

Fig. 7-4 Plots of organic nitrogen and ammonia nitrogen (in milli-grams per kilogram of dry weight) versus depth for a sediment core from the Santa Barbara Basin, California. (*Data are from Rittenberg et al., 1955.*)

expansion for $\exp[(-k/\omega)x]$ in Eq. (13).   Eliminating all terms higher than the first order, Eq. (13) is reduced to the linear expression

$$N = N_0 \left(1 - \frac{k}{\omega}x\right) \tag{20}$$

Differentiating Eqs. (19) and (20),

$$\frac{dc}{dN} = -\frac{\omega^2}{\omega^2 + D_s k} \tag{21}$$

$$\frac{dN}{dx} = -\frac{kN_0}{\omega} \tag{22}$$

Solving for D_s,

$$D_s = \frac{[(dc/dN) + 1]\,\omega N_0}{(dc/dN)(dN/dx)} \tag{23}$$

or

$$D_s = \frac{[(dc/dx) + (dN/dx)]\,\omega N_0}{(dc/dx)(dN/dx)} \tag{24}$$

The value for N_0 is not known but it must be between the limits 1300 mg/kg and 3500 mg/kg.   The latter value represents the total organic nitrogen at $x = 0$.   The lesser value represents the loss of organic nitrogen (presumably due to ammonia formation) down to 200 cm depth.   Since the decomposable organic nitrogen must be less than the total organic nitrogen, an intermediate value is probable.   With this in mind, these limiting values plus values for dc/dx and dN/dx taken from Fig. 7-4 can be combined with the value $\omega = 0.08$ cm/yr* for the Santa Barbara Basin (Emery, 1960) to obtain limiting values for D_s.   Using Eq. (24) the result is

$$D_s < 1.0 \times 10^{-5} \text{ cm}^2/\text{sec}$$

$$D_s > 0.4 \times 10^{-5} \text{ cm}^2/\text{sec}$$

This range of values is reasonable for dissolved constituents in sediments (see Chapter 6) and suggests that the theoretical model is a fairly good representation of the steady-state diagenesis of nitrogen.

Most sediments undoubtedly do not represent the simplified situation of steady-state diagenesis.   In such situations models more complicated than the one discussed above would be required.   However, a steady-state model is still useful as a reference state to which other more complicated models can be compared.

Molecular nitrogen, N_2, may be formed in anaerobic sediments by the bacterial reduction of NO_3^-.   This process, known as denitrification, is thermodynamically predicted by Fig. 7-1 and can be represented, using carbohydrate, by the reaction

$$5CH_2O + 4NO_3^- \leftrightarrows 2N_2 + 4HCO_3^- + CO_2 + 3H_2O$$

Increases of dissolved N_2 in anaerobic waters as a result of bacterial activity have been reported by Richards and Benson (1961) for bottom waters of Norwegian fjords and the Cariaco Trench, and by Kriss (1963) for deeper waters of the Black Sea.   Kriss attributes the excess N_2 to the bacterial reduction of nitrate which results from earlier anaerobic ammonia oxidation. The NO_3^- is supposedly removed as fast as it is formed.   Anaerobic ammonia oxidation, however, has not been documented and anyhow would not be expected thermodynamically.

Removal of dissolved N_2 to form bacterial protein, a process known as fixation, may also occur in marine sediments.   The process is well documented for marine bacteria in the laboratory (Thimann, 1963).   If the bacterial protein is subsequently decomposed to ammonia, the overall process amounts to the reduction of $N_{2\,aq}$ to $NH_{3\,aq}$ and $NH_{4\,aq}^+$.   In this way N_2 reduction to ammonia, a process predicted by the thermodynamic considerations discussed earlier, may be accomplished in marine sediments.   However, ammonia formed in this manner would be far less abundant than that formed by the decomposition of originally deposited organic nitrogen compounds.   Organic

---

* The average value for ω, based both on radiocarbon dating and varve counts, is 0.16 cm/yr for the Santa Barbara Basin.   However, the core on which the data of Fig. 7-4 are based showed 50 percent shortening during coring so that the value of ω is accordingly reduced by one-half.

nitrogen in anaerobic sediments is generally over a thousand times more abundant than dissolved N_2 (and NO_3^-). Whether the net change of dissolved N_2 in a sediment is an increase or decrease depends upon the relative importance of denitrification and fixation. Little data on the concentration of N_2 in pore waters is available and much more is needed.   The preliminary work of Reeburgh (1967) suggests that there is little change in pore water N_2 which can be attributed to biological processes.

DIAGENESIS OF SULFUR

Sulfur is added to sediments in two forms: organic sulfur compounds and dissolved SO_4^{--}. Subsequent diagenesis is strongly dependent upon the presence or absence of dissolved O_2. Under aerobic conditions organic sulfur compounds are simply oxidized to sulfate and added to the original sulfate which undergoes no change. Under anaerobic conditions sulfate-reducing bacteria can reduce sulfate to H_2S and HS^- and additional sulfide may be added by the bacterial degradation of sulfur in organic matter. The dissolved sulfide subsequently reacts with iron minerals to form iron sulfides (see Chapter 10) or due to migration comes into contact with dissolved O_2 and is oxidized either inorganically or by bacteria to elemental sulfur and sulfate. The end products of anaerobic diagenesis are pyrite or marcasite and in rarer instances elemental sulfur.

SULFATE REDUCTION

The source of H_2S in sediments depends upon the original concentration of SO_4^{--} in the pore waters. In some lakes which contain very little SO_4^{--}, the H_2S in organic-rich sediments may come for the most part from the breakdown of protein and amino acids, such as methionine, which contains sulfur. In most sediments, however, dissolved sulfide is the product of bacterial sulfate reduction. This is corroborated by at least two lines of evidence. First, there is a reasonably good correlation between H_2S in organic sediments and SO_4^{--} in associated lake waters. In sea water, which has a relatively high sulfate concentration, H_2S formation in organic sediments is extremely common. Secondly, in many marine sediments the concentration of pyritic sulfur (which forms from H_2S) is far greater than that which could have been originally supplied as organic sulfur (see Chapter 10).

As stated earlier bacterial sulfate reduction is a process by which organic carbon is oxidized under anaerobic conditions. The general reaction for carbohydrates is

$$2CH_2O + SO_4^{--} \rightarrow H_2S + 2HCO_3^-$$

The bacteria which reduce sulfate to H_2S (*Desulfovibrio*) are normally heterotrophic and are obligate anaerobes. The latter explains the absence of dissimilative (H_2S-forming) sulfate reduction in aerated waters. Of course, assimilative sulfate reduction to protein sulfur by higher plants and bacteria in general takes place during biosynthesis in both aerobic and anaerobic waters. Sulfate-reducing bacteria have been found to produce H_2S in the laboratory and in nature using a wide variety of organic sources as well as H_2 gas. The organic sources include hydrocarbons, fatty acids, carbohydrates, and amino acids (Thimann, 1963). The bacteria are also easily adapted to a wide range of temperatures and pressures as shown by their presence in deep sea sediments, Antarctic muds, and thermal waters.

The rate of nonbiological reduction of SO_4^{--} to H_2S by geologically reasonable reducing agents at room temperature is immeasurably slow. Thus, bacteria serve as necessary catalysts. Bacterial reduction is dependent, among other things, upon the availability of oxidizable organic matter. If so, the metabolic rate can be represented as

$$\frac{dC}{dt} = -LkG \tag{25}$$

where C = concentration of sulfate

t = time

G = concentration of organic carbon available for sulfate reduction

k = bacterial rate constant

L = stoichiometric coefficient relating the number of atoms of sulfate reduced per atom of carbon oxidized (for the CH_2O reaction $L = \frac{1}{2}$)

If bacterial reduction plus diffusion are the only processes affecting sulfate concentration, as is the case in most sediments, the general diagenetic equation (see Chapter 6) becomes

$$\frac{\partial C}{\partial t} = D_s \frac{\partial^2 C}{\partial x^2} - LkG - \omega \frac{\partial C}{\partial x} \tag{26}$$

In order to solve this equation an expression for G is necessary.

If the organic carbon source is present mostly in a nondiffusable form, its concentration is a sole function of bacterial decay rate. If the decay rate is first order, i.e.,

$$\frac{dG}{dt} = -k^1 G \tag{27}$$

substitution in the diagenetic equation for G yields

$$\frac{\partial G}{\partial t} = -k^1 G - \omega \frac{\partial G}{\partial x} \tag{28}$$

If steady-state diagenesis is assumed

$$\frac{\partial G}{\partial x} = \frac{-k^1}{\omega} G \tag{29}$$

and

$$G = G_0 \exp\left[-\left(\frac{k^1}{\omega}\right)x\right] \tag{30}$$

To be able to determine G_0, it is further assumed that most of the decreas in available organic carbon with depth is due to sulfate reduction.   If so

$$k^1 = k$$

Thus, for steady-state diagenesis

$$D_s \frac{\partial^2 C}{\partial x^2} - LkG_0 \exp\left[-\left(\frac{k}{\omega}\right)x\right] - \omega \frac{\partial C}{\partial x} = 0 \tag{31}$$

Solution of Eq. (31) for the boundary conditions

$x = 0$ $C = C_0$ (concentration in overlying water)

$x \to \infty$ $C \to C_\infty$ (concentration of SO_4^{--}, after reduction ceases)

yields

$$C = \frac{\omega^2 LG_0}{\omega^2 + D_s k} \exp\left[-\left(\frac{k}{\omega}\right)x\right] + C_\infty \tag{32}$$

where

$$\frac{\omega^2 LG_0}{\omega^2 + D_s k} = C_0 - C_\infty \tag{33}$$

For steady-state sulfate reduction the shape of the curve of C versus x would thus be expected to be concave downward and approach an asymptotic value. This is a consequence of the loss of utilizable organic matter with depth which should bring about an accompanying deceleration of sulfate reduction.

The steady-state model can be tested from measurements of a natural sediment which shows a homogeneous lithology and concave-down curves of sulfate and organic carbon versus depth.   If the rate of deposition is known and G_0 can be assumed to represent the difference between total organic carbon at $x = 0$ and the approximate value where $C \approx C_\infty$, curve fitting allows solution of the model for D_s or other parameters which can be determined independently.   This has been done by Berner (1964a) using data measured by Kaplan et al. (1963) for homogeneous finely laminated muds from the

Santa Barbara Basin off southern California.   Curve-fitting plus substitution for G_0 and for ω (as measured by C^{14} dating) enabled calculation of D_S

$$D_S = 0.3 \times 10^{-5} \text{ cm}^2/\text{sec} \tag{34}$$

This is a reasonable approximate value for the diffusion coefficient of a non-exchangeable anion in a fine-grained sediment (Chapter 6).

From curve-fitting, the value of k also was obtained and was used to calculate the rate of sulfate reduction near the sediment surface:

$$\frac{dC}{dt_{x=0}} = -LkG = -1.8 \times 10^{-14} \text{ moles } S/\text{cm}^3/\text{sec} \tag{35}$$

Kaplan et al. (1963), from the average concentration of pyrite in the upper 10 centimeters, the rate of deposition, and the assumption that sulfate is completely converted to pyrite, calculated the value

$$\frac{dC}{dt_{x=0}} = -6 \times 10^{-14} \text{ mole } S/\text{cm}^3/\text{sec} \tag{36}$$

The order of magnitude agreement is thought to be good considering all the assumptions made in calculating each value.   From the above it is concluded that Eq. (32) constitutes a valid model for the steady-state diagenesis of sulfate.

Non-steady-state sulfate diagenesis is probably much more common. This is shown by the curves of sulfate concentration versus depth found for H_2S-containing sediments in the Gulf of California (Berner, 1964b).   Of five sediment cores studied only one showed a concave-down, asymptotic curve. The others showed linear to concave-up curves in the upper meter and the absence of H_2S in the top 20 cm.   This indicates a decrease with time in the sulfate reduction rate at the sediment-water interface and, thus, non-steady-state diagenesis.

SULFIDE OXIDATION

Once H_2S is formed in an anaerobic sediment or bottom water it may react with dissolved O_2 in the overlying aerobic water.   This reaction may result from interdiffusion of sulfide and oxygen at a sharp interface or be brought about by the mixing of oxygenated water with sulfidic sediment as a result of storm stirring or the burrowing activities of benthonic organisms.   The reaction is often mediated by bacteria such as the thiobacilli although it occurs rapidly in the absence of microorganisms.   Bacterial oxidation results in the formation of elemental sulfur and/or sulfate.   Some forms, such as *Beggiatoa*, secrete elemental sulfur in the form of tiny globules.   The globules are used as an energy reserve and oxidized to SO_4^{--} when the concentration of H_2S drops.   Accumulation of white sulfur globules to form a thin bacterial "plate" at the interface between waters containing H_2S and O_2 is a common phenomenon in meromictic lakes.   In the presence of light, H_2S is also oxidized to sulfur and sulfate *anaerobically* by the photosynthetic purple and green bacteria. Kriss (1963) has described bacterial elemental sulfur formation at the bottom of the Black Sea, which is also anaerobic but not photosynthetic. However, this process has not been documented in the laboratory (e.g., Thimann, 1963) and must be considered as a speculation. More work is needed to check for the possibility of nonphotosynthetic, anaerobic sulfur formation because of its importance to the problem of the origin of sedimentary pyrite (see Chapter 10).

BIOGEOCHEMICAL CYCLE OF CARBON, NITROGEN, AND SULFUR

The above discussion can be summarized in the form of a diagram, Fig. 7-5, which illustrates the combined biogeochemical cycles of carbon, nitrogen, and sulfur. Because the diagram is intended primarily to show the major chemical species formed in natural waters and sediments, many other reactions, both biological and nonbiological, are omitted for the sake of simplicity.

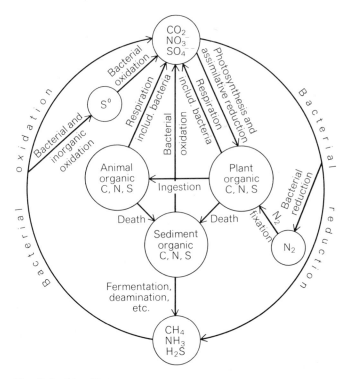

Fig. 7-5 Simplified biogeochemical cycle of carbon, nitrogen, and sulfur emphasizing predominant chemical species in natural waters and sediments.

REFERENCES

Allen, P. L., and Hickling, A., 1957, Electrochemistry of sulphur, I., Overpotential in the discharge of the sulphide ion, *Trans. Faraday Soc.*, v. 53, pp. 1626–1635.

Berner, R. A., 1963, Electrode studies of hydrogen sulfide in marine sediments, *Geochim. et Cosmochim. Acta*, v. 27, pp. 563–575.

———, 1964a, An idealized model of dissolved sulfate distribution in recent sediments, *Geochim. et Cosmochim. Acta*, v. 28, pp. 1497–1503.

———, 1964b, Distribution and diagenesis of sulfur in some sediments from the Gulf of California, *Marine Geology*, v. 1, pp. 117–140.

———, 1966, Chemical diagenesis of some modern carbonate sediments, *Am. Jour. Sci.*, v. 264, pp. 1–36.

———, 1968, Calcium carbonate concretions formed by the decomposition of organic matter, *Science*, v. 159, pp. 195–197.

Bockris, J. O'M., and Oldfield, L. F., 1955, The oxidation-reduction reactions of hydrogen peroxide at inert metal electrodes and mercury cathodes, *Trans. Faraday Soc.*, v. 51, pp. 249–259.

Doyle, R. W., 1967, Eh and thermodynamic equilibrium in environments containing dissolved ferrous ion, Ph.D. Dissertation, Yale University, 100 p.

Eglinton, G., and Murphy, M. T. J., 1969, *Organic geochemistry, methods and results*, Springer-Verlag, Berlin, 828 p.

Emery, K. O., 1960, *The sea off southern California*, Wiley, New York, 366 p.

———, and Rittenberg, S. C., 1952, Early diagenesis of California basin sediments in relation to origin of oil, *Am. Assoc. Petroleum Geologists Bull.*, v. 36, pp. 735–806.

Feely, H. W., and Kulp, J. L., 1957, Origin of Gulf Coast saltdome sulphur deposits, *Am. Assoc. Petroleum Geologists Bull.*, v. 41, pp. 1803–1853.

Garrels, R. M., and Christ, C. L., 1965, *Solutions, minerals, and equilibria*, Harper, New York, 450 p.

Kaplan, I. R., Emery, K. O., and Rittenberg, S. C., 1963, The distribution and isotopic abundance of sulphur in recent marine sediments off southern California, *Geochim. et Cosmochim. Acta*, v. 27, pp. 297–331.

Krause, H. R., 1959, Biochemische Untersuchungen über den postmortalen Abbau von totem Plankton unter aeroben und anaeroben Bedingungen, *Archiv. fur Hydrobiol.*, v. 24, pp. 297–337.

Kriss, A. E., 1963, *Marine microbiology*, Oliver & Boyd, Edinburgh, 536 p.

McCarty, P. L., 1964, The methane fermentation, in Heukelekian, H., and Dondero, N. C., *Principles and applications in aquatic microbiology*, Wiley, New York, pp. 314–343.

Presley, B. J., and Kaplan, I. R., 1968, Changes in dissolved sulfate, calcium, and carbonate from interstitial water of near-shore sediments, *Geochim. et Cosmochim. Acta*, v. 32, pp. 1037–1049.

Reeburgh, W. S., 1967, Measurement of gases in sediments, Ph.D. Thesis, Johns Hopkins University, 93 p.

Richards, F. A., 1965, "Anoxic basins and fjords," in Riley, J. P. and Skirrow, G., *Chemical Oceanography*, I, Academic, New York, pp. 611–645.

———, and Benson, B. B., 1961, Nitrogen/argon and nitrogen isotope ratios in two anaerobic environments, the Cariaco Trench in the Caribbean Sea and Dramsfjord, Norway, *Deep-Sea Research*, v. 7, pp. 254–264.

Rittenberg, S. C., Emery, K. O., and Orr, W. L., 1955, Regeneration of nutrients in sediments of marine basins, *Deep-Sea Research*, v. 3, pp. 23–45.

Sato, M., 1960, Oxidation of sulfide ore bodies, 1.   Geochemical environments in terms of Eh and pH, *Econ. Geol.*, v. 55, pp. 928–961.

Stumm, W., 1966, Redox potential as an environmental parameter; conceptual significance and operational limitation, *Third Internat. Conf. on Water Pollution Research*, Paper No. 13, pp. 1–16.

Thayer, L. A., 1931, Bacterial genesis of hydrocarbons from fatty acids, *Am. Assoc. Petroleum Geologists Bull.*, v. 15, pp. 441–453.

Thimann, K. V., 1963, *The life of bacteria*, 2d ed., Macmillan, New York, 775 p.

Thorstenson, D. C., 1970, Equilibrium distribution of small organic molecules in natural waters, *Geochim. et Cosmochim. Acta*, v. 34, pp. 745–770.

Voge, H., 1939, Exchange reactions with radio sulfur, *Jour. Am. Chem. Soc.*, v. 61, pp. 1032–1035.

Wood, E. J. F., 1965, *Marine microbial ecology*, Chapman & Hall, London, 243 p.

ZoBell, C. E., 1946, *Marine microbiology*, Chronica Botanica Co., Waltham, Mass., 240 p.

8
Diagenesis of
Ca–Mg Carbonates

Although many sedimentary minerals exist in the system $CaO–MgO–CO_2–H_2O$, this chapter will be concerned only with aragonite, calcite, and dolomite. Together these three constitute over 99 percent of all Ca–Mg carbonate minerals in sediments and sedimentary rocks. The composition of sedimentary aragonite is essentially pure $CaCO_3$. Sedimentary calcite is best represented by the formula $(Ca_{1-x}Mg_x)CO_3$, where $0 < x < 0.25$. Calcite with $x > 0.10$ is referred to as high-magnesium calcite, whereas that with $x < 0.05$ is referred to as low-magnesium calcite. Sedimentary dolomite can be represented by the formula $Ca_{1+x}(Mg, Fe)_{1-x}(CO_3)_2$, with $0 \leqq x < 0.12$. Iron substitution in most dolomite is low and will be neglected for the present purposes.

Most ancient carbonate rocks, on the basis of fossil or other evidence, were deposited in relatively shallow sea water. The minerals are almost entirely low-magnesium calcite and dolomite. On the other hand, the principal minerals in modern shallow-water marine carbonate sediments are high-magnesium calcite and aragonite which are derived for the most part

138

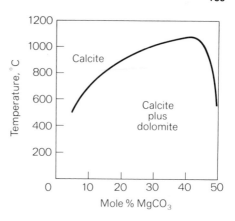

Fig. 8-1 Dolomite–magnesian calcite equilibrium relations as a function of temperature and composition. Note that at low temperature, magnesian calcites with greater than 5 mole percent $MgCO_3$ are unstable relative to dolomite plus calcites lower in $MgCO_3$. (*After Goldsmith and Heard, 1961.*)

from the disaggregation of marine skeletal materials.   Species which presently secrete aragonite or high-magnesium calcite are found in Pleistocene and Tertiary rocks as fossils consisting mainly of low-magnesium calcite.   If ancient carbonate rocks and fossils were originally of the same mineral composition as is found in shallow-water marine carbonate sediments, then large-scale mineralogical changes must take place as a result of diagenesis.

Diagenetic mineral changes are in agreement with predictions based on thermodynamic data.   For the reaction

$$CaCO_{3\ aragonite} \;\rightarrow\; CaCO_{3\ calcite}$$

$\Delta F°$ at all sedimentary pressures and temperatures based on many studies is definitely less than zero, on the order of minus a few hundred calories.   For the dolomite exsolution reaction

$$(1 - 2y)(Ca_{1-x}Mg_x)CO_3 \;\rightarrow\; (1 - 2x)(Ca_{1-y}Mg_y)CO_3 +$$
$$(x - y)CaMg(CO_3)_{2\ dolomite}$$

where $x > y$. $\Delta F°$ at sedimentary pressures and temperatures is definitely less than zero for $x \geqq 0.05$ as based on the phase-equilibria studies of Goldsmith and Heard (1961) (see Fig. 8-1).   Therefore, the more stable phases in sedimentary rocks are dolomite and low-magnesium calcite, and these are what form during diagenesis.

LIMESTONE FORMATION

The term *limestone* is here used to refer to a rock consisting mainly of low-magnesium calcite.   The overall diagenetic reactions in limestone formation are

A. $CaCO_{3\ aragonite} \;\rightarrow\; CaCO_{3\ calcite}$

B. $x\text{H}_{aq}^+ + (\text{Ca}_{1-x}\text{Mg}_x)\text{CO}_{3\,\text{calcite}} \rightarrow x\text{Mg}_{aq}^{++} + x\text{HCO}_{3\,aq}^- +$

$$(1 - x)\text{CaCO}_{3\,\text{calcite}}$$

C. $\text{Ca}_{aq}^{++} + \text{MgCO}_{3\,\text{calcite}} \rightarrow \text{CaCO}_{3\,\text{calcite}} + \text{Mg}_{aq}^{++}$

Reaction A is the transformation of aragonite to calcite, reaction B is the loss to solution of MgCO_3 in high-magnesium calcite, and reaction C is the exchange of Ca^{++} for Mg^{++} in high-magnesium calcite.

EXPERIMENTAL AND THEORETICAL STUDIES

Studies of reaction A have been extensive.   The recent work of Bischoff and Fyfe (1968) and Bischoff (1968) along with results of earlier studies (which are summarized in these papers) enable the following generalizations:

1. At sedimentary temperatures reaction A proceeds by the simultaneous solution of aragonite and precipitation of calcite.   This is much faster than solid-state transformation, which requires high temperatures to achieve appreciable reaction over geologic times.
2. Dissolved magnesium, and to a much lesser extent dissolved sulfate, greatly inhibit the rate of transformation of aragonite to calcite.   Only 5 ppm Mg_{aq}^{++} causes a four-fold reduction in the rate of transformation over that in Mg-free water.   At sea-water concentrations of Mg^{++} ($\approx 0.05\,m$) no observable transformation to low-Mg calcite has been achieved in the laboratory at sedimentary temperatures.
3. Nucleation of calcite is heterogeneous with aragonite acting as the nucleating agent.
4. Dissolved Mg inhibits transformation because of its uptake by calcite embryos and consequent destabilization which, if Mg uptake is high enough, prohibits nucleation and growth of low-magnesium calcite. Actual Mg uptake is demonstrated by the work of Bischoff (1968). Dissolved Mg does not prevent transformation by the armoring and blocking of dissolution of aragonite as suggested by Taft (1967).

Additional experimental work is in agreement with the conclusions stated above.   Berner (1966b) and de Groot and Duyvis (1966) have demonstrated that high- and low-Mg calcite crystals in Mg^{++}-rich solutions concentrate magnesium at their surfaces.   The magnesium undergoes exchange for calcium via the reaction (Berner, 1966b)

$$\text{Ca}_{aq}^{++} + \text{Mg}_{surf} \leftrightharpoons \text{Mg}_{aq}^{++} + \text{Ca}_{surf}$$

Assuming ideal solution on the exchange sites ($\lambda = 1$)

$$K = \frac{a_{\text{Mg}^{++}}/a_{\text{Ca}^{++}}}{(X_{\text{Mg}}/X_{\text{Ca}})_{surf}} = 1.5 \tag{1}$$

The relative constancy of K found over large variations in $a_{Mg^{++}}/a_{Ca^{++}}$ indicates an essentially ideal substitution of Ca for Mg on the calcite surface. At the sea-water value of $a_{Mg^{++}}/a_{Ca^{++}}$ of ≈ 5.7, $(X_{Mg}/X_{Ca})_{surf} \approx 4$. Therefore, the surfaces are much higher in Mg than the interior of the crystals. The bulk value for even high-Mg calcite is only $(X_{Mg}/X_{Ca}) \approx 0.16$. The magnesium enrichment at the calcite surface may be analogous to magnesium enrichment in calcite embryos (which consist almost entirely of surface). Thus, the presence of calcite seed crystals in Mg^{++}-rich solutions does not bring about the heterogeneous nucleations of low-Mg calcite because of the "poisoning" of the surfaces of the seeds by Mg.

Bischoff and Fyfe (1968) fitted their rate data for calcite crystal growth from aragonite in solutions not containing Mg_{aq}^{++} (or $SO_{4\,aq}^{--}$) with the expression

$$W = kt^2 \tag{1a}$$

where W = weight percent calcite = $100(Wt_{calcite}/Wt_{calcite} + Wt_{aragonite})$

$\quad t$ = time

$\quad k$ = constant

Actually the data are fitted better by the expression (see Fig. 8-2)

$$W = kt^{\frac{3}{2}} \tag{2}$$

Since the final calcite crystals were all found to be of approximately the same size, it can be assumed, as pointed out in Chapter 2, that the empirical expression refers mainly to growth on a constant number of initial nuclei. If so, the $t^{\frac{3}{2}}$

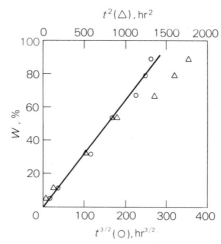

Fig. 8-2 Plot of w (wt. percent of calcite) versus time functions for representative data of Bischoff and Fyfe (1968). Note the better straight line fit with $t^{\frac{3}{2}}$ than with t^2.

dependency indicates diffusion-controlled growth. This can be seen as follows:

$$Wt_{calcite} = \frac{W}{100}(Wt_{calcite} + Wt_{aragonite}) \tag{3}$$

Also

$$Wt_{calcite} = \tfrac{4}{3}\pi r^3 \, dn \tag{4a}$$

$$Wt_{calcite} + Wt_{aragonite} = \tfrac{4}{3}\pi r_f^3 \, dn \tag{4b}$$

where r = radius of a sphere having the same volume as each calcite crystal

r_f = radius of calcite crystals when $W = 100$ percent

d = density

n = number of crystals of calcite

From Eqs. (2), (3), and (4)

$$r = k't^{\frac{1}{2}} \tag{5}$$

where $k' = r_f(k/100)^{\frac{1}{3}}$. Equation (5) can be identified with the integrated form of Eq. (88) of Chapter 2, the expression for diffusion-controlled growth. If so, then

$$k' = [2vD(C_\infty - C_s)]^{\frac{1}{2}} \tag{6}$$

Saturation with aragonite throughout the calcite-growth experiments was assumed by Bischoff and Fyfe. In this case $C_\infty = C$ for aragonite equilibrium and $C_s = C$ for calcite equilibrium at the same temperature. For growth in 0.01 m NaCl solution the final calcite crystal size is shown to be $r_f \approx 20$ microns (Bischoff, 1966, plate II). The time necessary to reach the value of $r_f = 20$ microns is that corresponding to $W = 100$ percent. From the data of Bischoff and Fyfe this is 32 hours for 0.01 m NaCl solution at 100°C. From these values of r_f and t, D can be calculated via Eqs. (5) and (6) if v and $C_\infty - C_s$ are known. The value for the differential solubility of aragonite and calcite in 0.01 m NaCl solution at 100°C is

$$C_\infty - C_s \approx (K_m^{\frac{1}{2}}{}_{aragonite} - K_m^{\frac{1}{2}}{}_{calcite}) \approx 10^{-8} \text{ mole/cm}^3$$

Also

$$r = 2 \times 10^{-3} \text{ cm} = r_f$$

$$t = 1.15 \times 10^5 \text{ sec}$$

$$v = 37 \text{ cm}^3/\text{mole}$$

From these values

$$D = 4.7 \times 10^{-5} \text{ cm}^2/\text{sec} \tag{7}$$

This is in good agreement with values predicted for dissolved species at the temperature of the experiments, 100°C.

Acceleration of growth with increasing ionic strength in KCl or NaCl solution was also found by Bischoff and Fyfe (1968). This can be explained simply by an increase in the value of $C_{s\,aragonite} - C_{s\,calcite}$ as a result of the decrease in ion-activity coefficients with increasing ionic strength.

Reaction B, like reaction A, involves dissolution and the crystallization of low-magnesium calcite. In this case, however, Mg^{++} is left in solution. Reaction B has been demonstrated by the experimental work of Friedman (1964) and Berner (1967). In the latter study reaction of samples of a natural sediment rich in high-Mg calcite with carbonated distilled water over periods of 5 to 20 hours resulted in values of Mg^{++}/Ca^{++} in solution higher than Mg/Ca of the bulk sediment and of the included Mg calcite. A much lesser effect was found for rapid (≈ 2 minute) dissolution of the same samples by dilute HCl. Thus, incongruent dissolution takes place on the time scale of hours. Small percentages of the total sample were dissolved in this study, and thus only the surfaces of mineral grains were involved in recrystallization to low-Mg calcite. Continued reaction of high-Mg calcite with distilled water saturated with CO_2 over a period of months results in the formation of amounts of low-Mg calcite large enough to be identified by x-ray diffraction (Friedman, 1964). Reaction of the sediment samples discussed above with carbonated 0.1 m $MgCl_2$ solutions results in a lower pH and consequently lower contribution of dissolved HCO_3^- to solution, than that predicted from reaction with distilled water at the same P_{CO_2}. This indicates that Mg_{aq}^{++} inhibits reaction B as might be expected by a common-ion effect. However, what has been said earlier concerning the inhibition by Mg_{aq}^{++} of low-Mg calcite crystallization from aragonite is probably also applicable to crystallization from high-Mg calcite.

Reaction C can take place only in calcium-rich waters, such as sea water, and is probably unimportant as an overall diagenetic reaction because the source of calcium in limestone is most likely the original aragonite and high-Mg calcite which contain far more calcium than associated pore waters. Reaction C is of interest, however, when written as an equilibrium reaction for sea water. At equilibrium for reaction C

$$K = \frac{a_{Mg^{++}}\, a_{CaCO_3}^{calcite}}{a_{Ca^{++}}\, a_{MgCO_3}^{calcite}} \tag{8}$$

or

$$K = \frac{a_{Mg^{++}}(1 - X_{MgCO_3})\lambda_{CaCO_3}}{a_{Ca^{++}}\, X_{MgCO_3}\lambda_{MgCO_3}} \tag{9}$$

From K and known values of $a_{Ca^{++}}$ and $a_{Mg^{++}}$ in sea water the value of X_{MgCO_3} for equilibrium with sea water could be calculated if the values of λ were known.   Unfortunately, they are not and the value of ΔF° of formation for $MgCO_3$ (magnesite), from which K is calculated, is also not known accurately. Calculation of X_{MgCO_3}, however, can be made from the data of Lerman (1965). Lerman fitted a regular solution model to magnesian calcites in equilibrium with dolomite as measured at high temperatures by Goldsmith and Heard (1961).   Magnesian calcites were treated as solid solutions of $CaCO_3$ and $Ca_{0.5}Mg_{0.5}CO_3$.   Extrapolation of the fitted curves enabled Lerman to calculate values of λ_{CaCO_3} and $\lambda_{Ca_{0.5}Mg_{0.5}CO_3}$ as functions of X_{CaCO_3} at 25°C. Using Lerman's model, reaction C is rewritten as

$$0.5\ Ca_{aq}^{++} + Ca_{0.5}Mg_{0.5}CO_3\,_{\text{calcite}} \rightleftharpoons 0.5\ Mg_{aq}^{++} + CaCO_3\,_{\text{calcite}}$$

The standard free energy of formation for $Ca_{0.5}Mg_{0.5}CO_3$ is simply one-half that for stoichiometric dolomite.   The equilibrium constant based on the data of Appendix I is

$$K = 0.82 = \left(\frac{a_{Mg^{++}}}{a_{Ca^{++}}}\right)^{\frac{1}{2}} \frac{X_{CaCO_3}\lambda_{CaCO_3}}{(1 - X_{CaCO_3})\lambda_{Ca_{0.5}Mg_{0.5}CO_3}} \tag{10}$$

From Lerman's model at 25°C Eqs. (39) and (40) of Chapter 2 for a regular solution are (with $A = 5$ and $B = 1$)

$$\ln \lambda_{Ca_{0.5}Mg_{0.5}CO_3} = 5X_{CaCO_3}^2 + (3 - 4X_{CaCO_3})\,X_{CaCO_3}^2 \tag{11}$$

$$\ln \lambda_{CaCO_3} = 5(1 - X_{CaCO_3})^2 + (1 - 4X_{CaCO_3})(1 - X_{CaCO_3})^2 \tag{12}$$

From Chapter 3, for average sea water $(a_{Mg^{++}}/a_{Ca^{++}})^{\frac{1}{2}} = 2.4$ Substitution for λ and $a_{Mg^{++}}/a_{Ca^{++}}$ in Eq. (10) enables solution (by trial and error):

$$X_{CaCO_3} = 0.93$$

Thus, $X_{Ca_{0.5}Mg_{0.5}CO_3} = 0.07$ from which $X_{MgCO_3} = 0.035$, and calcite in equilibrium with sea water should contain 3.5 mole percent $MgCO_3$ according to the Lerman model.   Since this value is sensitive to errors in temperature extrapolation and the value of ΔF° for dolomite, it can only be considered approximate. However, it does indicate that low-magnesium calcite is the stable form of calcite in sea water, which is in accord with geologic data.

STUDIES OF NATURAL SEDIMENTS

From the experimental and theoretical work discussed above, the diagenetic crystallization of low-magnesium calcite would be expected to take place much more rapidly in fresh water low in dissolved Mg^{++} than in sea water which contains about 0.05 mole of Mg_{aq}^{++} per liter.   This is borne out by studies of natural carbonate sediments and sedimentary rocks.   In fine-grained (and therefore reactive) recent marine sediments, consisting of aragonite plus high-Mg calcite, evidence for the diagenetic formation of low-Mg calcite with depth, i.e., time, has not been found either by studies of mineralogy (Taft and

Harbaugh, 1964; Pilkey, 1964) or by analyses of pore waters for Mg_{aq}^{++} (Berner, 1966a). This is shown in Fig. 8-3. Aragonite skeletons of reef organisms buried in sea water for millions of years beneath several atolls in the Pacific Ocean show no mineralogical, textural, or isotopic evidence of recrystallization to low-Mg calcite (Gross and Tracey, 1966; Schlanger, 1963). In limited zones beneath the atolls where recrystallization to low-Mg calcite is found, there is evidence, as in the presence of land snail fossils (Schlanger, 1963) and in lowered O^{18}/O^{16} ratios (Gross and Tracey, 1966), of interaction with fresh water.

In contrast to submarine carbonate sediments, those that have been recently uplifted above sea level into the zone of fresh-water interaction show evidence of rapid transformation of aragonite and high-Mg calcite to low-Mg

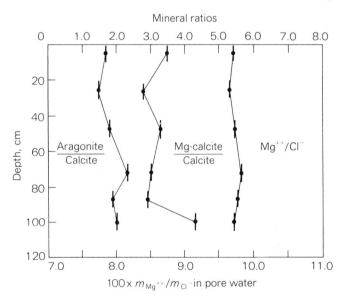

Fig. 8-3 Mineralogy and pore water Mg^{++}/Cl^- ratio versus depth for a core of fine-grained marine carbonate sediment from Florida Bay, Florida. The ratio of Mg^{++} to chloride ion is used in order to correct for variations in Mg^{++} concentration due to variations in salinity. Mineral ratios represent the ratio of integrated peak areas from x-ray diffractograms. Mg-calcite refers to a mixture of high-magnesium calcites showing a distinct mode at 14 mole percent $MgCO_3$. If low-magnesium calcite were being formed diagenetically, continuous decreases in the mineral ratios and a large increase in Mg^{++}/Cl^- with depth would be expected. [Dissolved Mg^{++} is a much more sensitive indicator of reaction B than is mineralogy (see Chapter I).] Lack of change in mineralogy and Mg^{++}/Cl^- indicates a lack of evidence for either dolomitization or low-magnesium calcite formation. (*After Berner, 1966a.*)

calcite. This is shown by petrographic evidence (Friedman, 1964; Matthews, 1968), isotopic evidence (Gross, 1964; Friedman, 1964) (see Fig. 8-4), and analyses of ground water (Berner, 1966a; Harris and Matthews, 1968). The rocks studied by the above workers are all Pleistocene or younger and are located on uplifted portions of the islands of Bermuda and Barbados. Ground-water analyses indicate incongruent solution of magnesium calcite via reaction B (Berner, 1966a) or of strontium-rich aragonite (Harris and Matthews, 1968) via an analogous reaction

$$x\text{H}_{aq}^+ + (\text{Ca}_{1-x}\text{Sr}_x)\text{CO}_{3\,\text{aragonite}} \rightarrow x\text{Sr}_{aq}^{++} + x\text{HCO}_{3\,aq}^- +$$
$$(1 - x)\text{CaCO}_{3\,\text{calcite}}$$

Since calcite accepts far less Sr^{++} in its crystal structure than does aragonite, the Sr^{++} accumulates in solution during the transformation of aragonite to calcite. Both Sr^{++} and Mg^{++} are subsequently removed from the sediment by ground-water flow.

The considerable lack of evidence for carbonate mineral transformation in modern submarine sediments should not be interpreted as indicating a total lack of CaCO_3 precipitation. As pointed out in Chapter 6 many studies have indicated that marine carbonate sediments may become cemented in sea water by calcium carbonate during early diagenesis. However, the cement is always aragonite or high-magnesium calcite and the resulting rock is not mineralogically the same as ancient limestones. Another factor is geologic time. Although fresh water, low in dissolved magnesium, greatly accelerates transformation, it is possible that over very long periods slow transformation to low-Mg calcite may also occur in sea water. Milliman (1966) has cited the occurrence of Miocene pelagic foraminifera in the deep ocean cemented by low-Mg calcite. The rocks are texturally and isotopically similar to associated younger material which is cemented by high-Mg calcite. By analogy with the younger rocks Milliman suggests that the low-Mg calcite has been derived from an original high-Mg calcite cement. Since there is little evidence that these rocks were ever in contact with fresh water, they may represent limestone that has formed directly, albeit slowly, from sea water.

Recent study of buried deep-sea sediments as a part of the JOIDES drilling program (Beall and Fischer, 1969) also suggests that low-magnesium calcite may form from sea water, given sufficient geologic time. Cretaceous calcarenites, consisting of shallow-water skeletal debris, which were re-deposited on the deep ocean floor and subsequently buried, are now free of aragonite and are extensively cemented by secondary calcite. Long slow transformation of aragonite and high-Mg calcite to low-Mg calcite in sea water may be an important diagenetic process for minerals which have never contacted fresh water, but much more evidence for this process is needed.

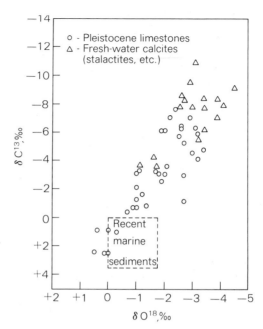

Fig. 8-4 Plot of oxygen and carbon isotopic composition for recent marine carbonate sediments, Pleistocene limestones, and fresh-water calcites of Bermuda.

$$\delta O^{18} = \left(\frac{O^{18}/O^{16}}{O^{18}/O_{std}^{16}} - 1\right) 1000$$

$$\delta C^{13} = \left(\frac{C^{13}/C^{12}}{C^{13}/C_{std}^{12}} - 1\right) 1000$$

The fresh-water calcites (stalactites, etc.) are all low-Mg calcite and exhibit δ values decidedly more negative than those for the recent marine sediments, which consist of aragonite and high-Mg calcite. This is expected in that the dissolved carbonate from which fresh-water calcite forms is isotopically enriched in C^{12} and O^{16} relative to dissolved carbonate in sea water. The Pleistocene limestones lie in an intermediate position between modern sediments and fresh-water calcite. This suggests a partial recrystallization to fresh-water calcite, from original aragonite and high-magnesium calcite. A good correlation between lowered δ values and degree of transformation to low-Mg calcite confirms this. Thus, the Pleistocene limestones have undergone, to varying degrees, subaerial fresh-water diagenesis. (*After Gross, 1964.*)

DOLOMITE FORMATION

The formation of sedimentary dolomite has been and to a lesser extent still is one of the great unsolved problems of sedimentary petrology. Although dolomite is a common constituent of ancient carbonate rocks it was not proved to be forming in modern sediments until relatively recently. Since 1957 several isolated occurrences of modern dolomite, all substantiated by x-ray diffraction and radiocarbon dating, have been reported. They include South Australian lagoons and lakes (Alderman and Skinner, 1957; Skinner, 1963; Skinner et al., 1963; von der Borch ,1965a), Deep Springs Lake, Calif. (Jones, 1961; Peterson et al., 1963), Persian Gulf mud flats (Wells, 1962; Curtis et al., 1963), a saline lake on the island of Bonaire, N.W.I. (Deffeyes et al., 1965), and supratidal flats of the Bahamas (Shinn et al., 1965). In every case the dolomite is associated with supersaline brines* and generally is accompanied by evaporite minerals. To the author's knowledge no modern dolomite has yet been found which is definitely forming from sea water of normal salinity.† The probable reason for the earlier lack of discovery of modern dolomite was that it was being sought in more typical marine sediments of average salinity where it does not form.

The lack of discovery of dolomite forming from sea water cannot be due to thermodynamic instability if recent measurements of its solubility product constant are accurate. For the reaction

$$CaMg(CO_3)_{2\,dolomite} \rightleftarrows Ca_{aq}^{++} + Mg_{aq}^{++} + 2CO_{3\,aq}^{--}$$

measurements summarized by Hsu (1967) and Langmuir (1964) indicate that at 25°C

$$K = 10^{-17.0} = a_{Ca^{++}}\,a_{Mg^{++}}\,a_{CO_3}^2{}^{--} \tag{13}$$

From the activities given in Chapter 3, the IAP for sea water is

$$IAP = 10^{-15.0} \tag{14}$$

Therefore, sea water is greatly supersaturated with dolomite. In addition dolomite is the most stable Ca-Mg carbonate in sea water. This can be seen from the reaction

$$Ca_{aq}^{++} + CaMg(CO_3)_{2\,dolomite} \rightleftarrows Mg_{aq}^{++} + 2CaCO_{3\,calcite} \tag{15}$$

---

* Supersalinity, however, is not necessary for dolomite formation at low temperatures since it has been found forming from nonsaline but alkaline spring waters (Barnes and O'Neil, 1969).

† The finding of suspended dolomite in South Australian waters of relatively normal marine salinity by Alderman and Skinner (1957) and Skinner (1963) is no assurance that it was forming from these waters. The salinity of the water where the dolomite is found fluctuates widely throughout the year, and the presence of dolomite in these waters at any one time can be readily ascribed to wind or biological stirring of the bottom sediments (von der Borch, 1965a).

where $K = 0.67 = a_{Mg^{++}}/a_{Ca^{++}}$.   In sea water $a_{Mg^{++}}/a_{Ca^{++}}$, from Chapter 3, is 5.7.   Therefore, dolomite is more stable than calcite (and thus aragonite) in sea water.   The reason why dolomite is not found forming in sea water must be due to some kinetic inhibition mechanism.

The chemistry of the formation of dolomite at low temperatures is very complex.   For one thing, it has never been synthesized at sedimentary temperatures.   The many reports in the literature of dolomite synthesis at low temperatures have incorrectly used the term dolomite for a ~50 mole percent magnesian calcite whose major x-ray diffraction peaks match those of dolomite.   Unless long-range Ca-Mg ordering can be demonstrated by the presence of at least faint traces of the principal-ordering x-ray reflections, then the material should not be called dolomite.   True dolomite is defined as having the $R\bar{3}$ space group as a result of ordering of Ca and Mg into alternating crystallographic planes.   Deficient ordering and a nonstoichiometric Ca-rich composition are commonly encountered in sedimentary dolomites (Goldsmith and Graf, 1958), but evidence for Ca-Mg ordering is still present, and this is not the case for the low-temperature synthetic materials which are best considered as magnesian calcites (Glover and Sippel, 1967).   The author has attempted to measure the solubility of a reasonably well-crystallized sample of synthetic "pseudo-dolomite" but found it to be so unstable that it incongruently dissolved to form aragonite plus Mg^{++} in solution.   This is not the case for *unground* samples of natural sedimentary dolomite, which exhibit congruent solubility and a pK of 17.0 (Hsu, 1967; Berner, 1967).

One of the basic questions in dolomite chemistry is whether true dolomite *can* form at low temperature by direct crystallization from aqueous solution.   Degens and Epstein (1964), based on a presumed identity of oxygen isotopic composition between coexisting dolomite and calcite in recent sediments, have stated that sedimentary dolomite forms by the solid-state replacement of Ca^{++} ions in $CaCO_3$ by Mg^{++} ions derived from aqueous solution.   This mechanism would preclude simple crystal growth as a dolomite-forming process.   However, several of their samples, based on original literature description, constitute detrital mixtures, and a recent study (Clayton et al., 1968b) has shown that South Australian recent dolomite (probably their best sample) is *not* isotopically identical to coexisting calcite.

Peterson et al. (1966) have advanced a solid-state replacement model for dolomite forming in Deep Springs Lake, California.   In this study it was found that the dolomite sediment, when subjected to successive periods of leaching by acid, showed Ca/Mg much higher in the first leachates than that expected for dolomite.   They interpreted this to mean that the dolomite crystals in the sediment are coated at their surfaces by a calcium-rich phase, probably magnesian calcite.   The crystals start to grow as calcite, and after burial in the crystal to a depth of a few atomic layers, Ca^{++} ions are replaced by Mg^{++} ions, diffusing through the overlying atomic layers to form dolomite.   The short distance of diffusion and presumed higher diffusivity through surface atomic layers would enhance solid diffusion, which is otherwise an extremely slow process at sedimentary temperatures.   In this manner the final dolomite crystals would have the oxygen isotopic composition predicted for calcite.   Clayton et al. (1968a) have studied Deep Spring Lake dolomitic sediment in great detail using x-ray diffraction, successive leachings by acid of different-size fractions, and oxygen and carbon isotopic analysis of each leachate sample.   Their results indicate that (1) the oxygen isotopic composition of the earliest leachates is distinctly different from the bulk of the dolomite crystals; (2) the bulk of the dolomite crystals does not exhibit an oxygen isotopic composition expected for calcite forming from the interstitial water of the sediment; (3) minute traces of calcite, detectable by x-ray diffraction, are found in the most dolomite-rich sediment samples; (4) calcite in nearby areas has an isotopic composition similar to the earliest leached material in the dolomitic sediment.   From these results it was concluded that the Ca-rich leachates found in the study by Peterson et al. came from traces of calcite (which dissolves in acid faster than dolomite) and that dolomite can form by simple crystal growth from aqueous solution and can have an isotopic composition different from that of calcite formed from the same solution.   The latter conclusion is in agreement with predictions based on studies of dolomite-calcite isotope fractionation at higher temperatures.

The rate of growth of the dolomite crystals in Deep Spring Lake has been measured by Peterson et al. (1963).   Various size fractions of the dolomite crystals were dated using C^{14} and were found to exhibit a linear relation between average C^{14} age and size which extrapolates to zero age at zero size.   The linear relation between age and size is also demonstrated by the leaching experiments of Peterson et al. (1966).   In order to obtain the true growth rate, correction has to be made for the fact that each crystal exhibits a spectrum of ages ranging from zero at its outer surface to a maximum at its core where crystal growth first began.   The average or measured value of C^{14}/C^{12} for a crystal is

$$\bar{Y} = \frac{\int_0^V Y(t)\,dV}{\int_0^V dV} \tag{16}$$

where $Y = C^{14}/C^{12}$ ratio at any point in a crystal

$\bar{Y}$ = average C^{14}/C^{12} ratio for the crystal

V = volume of crystal

Since the dolomite crystals occur as rhombohedra, they can be approximated as cubes so that

$$dV = 3x^2\,dx \tag{17}$$

where x = edge size of a crystal during growth. Also, since the rate of growth is linear

$$x = k(T - t) \tag{18}$$

$$dx = -k\,dt \tag{19}$$

where k = growth rate

t = age of volume element, dV

T = total time since nucleation

The integrated radioactive decay equation for C^{14} (since it follows first-order kinetics) is

$$Y = Y_0 \exp(-\lambda t) \tag{20}$$

where λ = decay constant = k_1

$Y_0 = Y$ at $t = 0$

Substituting Eqs. (17) to (20) in Eq. (16)

$$\bar{Y} = \frac{\int_0^T 3Y_0 k^3 \exp(-\lambda t)(T - t)^2\,dt}{\int_0^T 3k^3(T - t)^2\,dt} \tag{21}$$

where T = time for growth of entire crystal.

Direct integration of Eq. (21) results in a complicated expression, but a reasonable simplification may be made. Since the ages found (< 2500 years) are considerably less than the half life of C^{14} (~ 5000 years), the expression $\exp(-\lambda t)$ can be approximated by $1 - \lambda t$ without introducing appreciable error. Therefore, Eq. (21) simplifies to

$$\bar{Y} = \frac{Y_0 \int_0^T (T - t)^2(1 - \lambda t)\,dt}{\int_0^T (T - t)^2\,dt} \tag{22}$$

Upon integration

$$\bar{Y} = Y_0[1 - \tfrac{1}{4}\lambda T] \tag{23}$$

Since (by approximation)

$$\bar{Y} = Y_0[1 - \lambda \bar{t}] \tag{24}$$

where $\bar{t}$ = average or measured age

$$T = 4\bar{t} \tag{25}$$

This means that the total time since growth is four times the measured C^{14} age for crystals of a given size. A plot of T versus X (present measured size of crystals) results in

$$\frac{dX}{dT} \approx 500 \text{ Å}/10^3 \text{ years} \tag{26}$$

Hundreds of angstroms per thousand years is an extremely slow growth rate and the linear relationship indicates a polynuclear growth mechanism (Chapter 2). If growth is polynuclear and extremely slow, the rates of *both* surface nucleation and surface growth spreading must be extremely slow. Therefore, surface spreading must not be diffusion-controlled; otherwise the growth mechanism would be mononuclear. The slow surface nucleation and growth probably are due to the necessarily slow ordering of Ca and Mg ions caused by relatively small differences in binding energy of Ca^{++} and Mg^{++} at a given surface crystallographic site. Much faster growth, as demonstrated in the laboratory, results in the formation of metastable magnesium calcite of dolomitic composition which does not exhibit long-range cation ordering. Thus, the probable reason why ordered dolomite has not been synthesized at low temperature is the necessarily slow growth rate required for cation ordering. It is interesting that no evidence exists that any organism has ever secreted dolomite as skeletal material, even though it is the most stable Ca-Mg carbonate in sea water. Apparently organisms have not evolved enzyme systems which sufficiently catalyze cation ordering so that dolomite can be grown biosynthetically. No living organism, including man, has as yet successfully precipitated true dolomite under sedimentary conditions!

The apparent correlation of dolomite formation in recent sediments with high salinity and its lack of formation from ordinary sea water has not yet been explained on a comprehensive theoretical basis. Deffeyes et al. (1965) suggest that a high Mg^{++}/Ca^{++} ratio, which commonly occurs in supersaline waters as a result of evaporative concentration of Mg^{++} accompanied by precipitation of Ca^{++} as gypsum, promotes dolomite formation. Von der Borch (1965a) presents data for South Australian carbonate sediments which show a crude correlation between dolomite occurrences and higher ratios of Mg^{++}/Ca^{++} in associated waters. Von der Borch also points out that a better correlation is found between elevated pH and dolomite formation, with dolomite occurring in lakes where the maximum pH, during the annual cycle of variation, reaches values as high as 10.2. Table 8-1 illustrates the correlation between carbonate mineralogy and pH. (Note the absence of dolomite at the highest Mg^{++}/Ca^{++} ratio.) Values of Mg^{++}/Ca^{++} and pH for waters associated with other dolomite occurrences vary considerably and no simple relationship including all occurrences is apparent. One major problem is that in most of the dolomite localities the associated waters fluctuate widely in

Table 8-1   Mineralogy, Mg^{++}/Ca^{++} ratios in associated waters, and maximum pH values for South Australian carbonate environments. [After von der Borch (1965a)]

Locality	Surface carbonate	Maximum pH	Molar Mg^{++}/Ca^{++} at maximum pH
Permanent Coorong Lagoon	Aragonite plus Mg-calcite	8.2–8.5	4.0–6.8
Main Lake Chain	"Dolomite" (non-ordered) plus Mg-calcite	8.5–9.1	6.8–10.2
Lakes marginal to ephemeral Coorong Lagoon	Ordered dolomite	9.7–10.2	13.5–15.5
	Dolomite plus magnesite	9.4–10.2	27
	Aragonite plus hydromagnesite	8.9–9.0	34

salinity and Mg^{++}/Ca^{++} with time.   A water sampled at any given time in contact with dolomite may not represent the water from which it is forming.

Dolomites that have formed from hypersaline brines need not be accompanied by evaporite minerals.   This is illustrated by the South Australian lakes, some of which dry up completely each year.   The salt and gypsum precipitated during drying are redissolved when water adds to the lakes during the rainy season.   By contrast dolomite that has formed from brine is not dissolved when the water becomes less saline.   To maintain salt balance, dissolved salts are lost during the dry season by downward and outward flow of highly concentrated ground water to the nearby sea (von der Borch, 1965b). The subsurface return to the open ocean of supersaline brines formed by evaporation in back-reef lagoons has been proposed as a process for dolomitizing reefs (Adams and Rhodes, 1960; Deffeyes et al., 1965).   This is shown in Fig. 8-5. The brine attains a high density and high Mg^{++} concentration during evaporation in the lagoon.   Due to its high density the brine then passes downward through the reef until it flows out into the open ocean on the other side of the reef, a process called reflux.   The high-Mg_{aq}^{++} concentration, combined with low-Ca_{aq}^{++} concentration due to gypsum precipitation in the lagoons, promotes dolomitization of $CaCO_3$ in the reef.   This hypothesis, although reasonable for the formation of dolomitic reef rock where there are associated evaporite deposits, encounters some difficulty in explaining the occurrence of evaporite-free dolomitized reefs.   This is especially true since the dolomitization reaction

$$Mg^{++} + 2CaCO_3 \;\rightarrow\; CaMg(CO_3)_2 + Ca^{++}$$

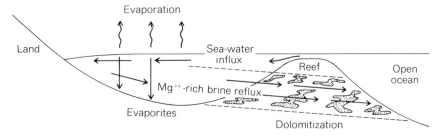

Fig. 8-5 Dolomitization of reefs by brine reflux. Dense, highly saline, Mg-rich brine returns to the sea by slow leakage through the permeable subsurface, and as a result older reef material becomes dolomitized. High ratios of Mg^{++}/Ca^{++} in the brine are brought about by evaporative concentration of Mg^{++} combined with $CaSO_4$ precipitation in the back-reef lagoon. Compare with Fig. 5-4. [*Adapted from Adams and Rhodes (1960) and Deffeyes et al. (1965).*]

generates calcium ions. If the downward refluxing brines were originally saturated with gypsum, subsurface gypsum precipitation should accompany dolomite formation because of the generation of additional calcium ions by dolomitization. One way out of the dilemma is to have later solution and flushing out of dissolved gypsum by less saline waters. The reflux hypothesis has not been proven in recent sediments and thus is still somewhat speculative. Deffeyes et al. (1965) did show brine reflux to be occurring on an island in the Netherlands West Indies, but they did *not* demonstrate that dolomite was forming in the subsurface from the refluxing brine.

Many ancient dolomites associated with limestones, but without evaporites, show evidence for deposition in very shallow water near shore-lines. A reasonable hypothesis for the formation of such shoreline dolomite has been proposed by Shinn et al. (1965). They have found that in modern supratidal mud flats dolomite is forming from the interaction of capillary brines with normal marine carbonate sediment. The supersaline brine results from the evaporation of sea water which, along with sediment, is thrown upon supratidal flats by infrequent unusually high tides. Subsequent tides (or rainfall) dissolve away any evaporite minerals which may precipitate during dolomitization. The dolomite forms a recrystallized rock which along with nearby normal marine carbonate sediments exhibits features similar to those found in many ancient dolomites and dolomitic limestones.

The origin of ancient dolomite rocks which are not associated with evaporites nor show evidence for nearshore deposition is more uncertain, because no modern examples are known. It is possible that during deep burial, supersaline brines which result from salt filtration due to Donnan phenomena (see Chapter 6) may have initially high-Mg^{++} concentrations which enable them to dolomitize limestone.   In this case there would be no evidence for evaporative hypersalinity or shoreline deposition.

If dolomite is more stable than calcite in sea water, given enough time, it should crystallize from calcite in submarine sediments.   Recent drilling in the Gulf of Mexico by the JOIDES organization (Beall and Fischer, 1969) has revealed the occurrence of considerable dolomite (up to 30 percent) in deep-sea sediments which have been buried in sea water for many millions of years. However, the dolomite occurs in association with clastics in turbidite sequences and is undoubtedly detrital.   That associated with pelagic calcium carbonate is probably to a large extent also detrital because the total mineralogy plots on a simple mixing curve between pure $CaCO_3$ (pelagic contribution) and average compositions for the turbidites.   Buried calcium carbonate sediments from the northeastern Atlantic Ocean contains much less dolomite (0 to 5 percent). Also, other occurrences of dolomite discovered by JOIDES deep-sea drilling appear to show a correlation with volcanic activity (C. C. von der Borch, personal communication).   Thus, at the time of writing, the author was not convinced of any evidence for areally extensive dolomitization of $CaCO_3$ by ordinary sea water over long periods of time.   According to the preliminary report from JOIDES Leg 1 (Beall and Fischer, 1969), long-term carbonate diagenesis, where it occurs in deep-sea sediments, is characterized by the formation of calcite, not dolomite.

REFERENCES

Adams, J. E., and Rhodes, M. L., 1960, Dolomitization by seepage refluxion, *Am. Assoc. Petroleum Geologists Bull.*, v. 44, pp. 1912–1920.

Alderman, A. R., and Skinner, H. C. W., 1957, Dolomite sedimentation in the southeast of South Australia, *Am. Jour. Sci.*, v. 255, pp. 561–567.

Barnes, I., and O'Neil, J. R., 1969, *The nature of secondary dolomite formation*, Abstract, EOS, *Trans. Am. Geophys. Union*, v. 50, p. 347.

Beall, A. O., and Fischer, A. G., 1969, Sedimentology, in *Initial reports of the Deep Sea Drilling Project*—JOIDES, Natl. Sci. Foundation, v. I, pp. 521–593.

Berner, R. A., 1966a, Chemical diagenesis of some modern carbonate sediments, *Am. Jour. Sci.*, v. 264, pp. 1–36.

———, 1966b, Diagenesis of carbonate sediments: Interaction of Mg^{++} in sea water with mineral grains, *Science*, v. 153, pp. 188–191.

———, 1967, Comparative dissolution characteristics of carbonate minerals in the presence and absence of aqueous magnesium ion, *Am. Jour. Sci.*, v. 265, p. 45–70.

Bischoff, J. L., 1966, Kinetics of crystallization of calcite and aragonite, Unpublished Ph.D. Dissertation, University of California at Berkeley.

———, 1968, Kinetics of calcite nucleation: Magnesium ion inhibition and ionic strength catalysis, *Jour. Geophys. Res.*, v. 73, pp. 3315–3321.

———, and Fyfe, W. S., 1968, Catalysis, inhibition, and the calcite-aragonite problem, *Am. Jour. Sci.*, v. 266, pp. 65–79.

Clayton, R. N., Jones, B. F., and Berner, R. A., 1968a, Isotope studies of dolomite formation under sedimentary conditions, *Geochim. et Cosmochim. Acta*, v. 32, pp. 415–432.

————, Skinner, H. C. W., Berner, R. A., and Rubinson, M., 1968b, Isotopic compositions of recent South Australian lagoonal carbonates, *Geochim. et Cosmochim. Acta*, v. 32, pp. 983–988.

Curtis, R. G., Evans, G., Kinsman, D. J. J., and Shearman, D. J., 1963, Association of dolomite and anhydrite in recent sediments of the Persian Gulf, *Nature*, v. 197, pp. 679–680.

Deffeyes, K. S., Lucia, F. J., and Weyl, P. K., 1965, Dolomitization of recent and Plio-Pleistocene sediments by marine evaporite waters on Bonaire, Netherlands Antilles, *Soc. Econ. Paleont. and Mineral., Spec. Publ. 13*, pp. 71–88.

Degens, E. T., and Epstein, S., 1964, Oxygen and carbon isotope ratios in coexisting calcites and dolomites from recent and ancient sediments, *Geochim. et Cosmochim. Acta*, v. 28, pp. 23–44.

De Groot, K., and Duyvis, E. M., 1966, Crystal form of precipitated calcium carbonate as influenced by adsorbed magnesium ions, *Nature*, v. 212, pp. 183–184.

Friedman, G. M., 1964, Early diagenesis and lithification in carbonate sediments, *Jour. Sed. Petrology*, v. 34, pp. 777–813.

Glover, E. D., and Sippel, R. F., 1967, Synthesis of magnesium calcites, *Geochim. et Cosmochim. Acta*, v. 31, pp. 603–614.

Goldsmith, J. R., and Graf, D. L., 1958, Structural and compositional variations in some natural dolomites, *Jour. Geol.*, v. 66, pp. 278–293.

Goldsmith, J. R., and Heard, H. C., 1961, Subsolidus phase relations in the system $CaCO_3$-$MgCO_3$, *Jour. Geol.*, v. 69, pp. 45–74.

Gross, M. G., 1964, Variations in the O^{18}/O^{16} and C^{13}/C^{12} ratios of diagenetically altered limestones of the Bermuda Islands, *Jour. Geol.*, v. 72, pp. 170–194.

————, and Tracey, J. R., 1966, Oxygen and carbon isotopic composition of limestones and dolomites, Bikini and Eniwetok Atolls, *Science*, v. 151, pp. 1082–1084.

Harris, W. H., and Matthews, R. K., 1968, Subaerial diagenesis of carbonate sediments: Efficiency of the solution-reprecipitation process, *Science*, v. 160, pp. 77–79.

Hsu, K. J., 1967, Chemistry of dolomite formation, in Chilingar, G. V., Bissell, H. J., and Fairbridge, R. W. (eds.), *Carbonate rocks*, Elsevier, Amsterdam, pp. 169–191.

Jones, B. F., 1961, Zoning of saline minerals at Deep Springs Lake, California, *U.S. Geol. Surv. Prof. Paper 421*, pp. B199–B202.

Langmuir, D., 1964, Stability of carbonates in the system CaO-MgO-CO_2-H_2O, Unpublished Ph.D. Thesis, Harvard University, Cambridge, Mass.

Lerman, A., 1965, Paleoecological problems of Mg and Sr in biogenic calcites in light of recent thermodynamic data, *Geochim. et Cosmochim. Acta*, v. 29, pp. 947–966.

Matthews, R. K., 1968, Carbonate diagenesis: equilibration of sedimentary mineralogy to the subaerial environment: Coral cap of Barbados, West Indies, *Jour. Sed. Petrology*, v. 38, pp. 1110–1119.

Milliman, J. D., 1966, Submarine lithification of carbonate sediments, *Science*, v. 153, pp. 994–997.

Peterson, M. N. A., Bien, G. S., and Berner, R. A., 1963, Radiocarbon studies of recent dolomite from Deep Spring Lake, California, *Jour. Geophys. Res.*, v. 68, pp. 6493–6505.

————, von der Borch, C. C., and Bien, G. S., 1966, Growth of dolomite crystals, *Am. Jour. Sci.*, v. 264, pp. 257–272.

Pilkey, O. H., 1964, Mineralogy of the fine fraction in certain carbonate cores, *Bull. Mar. Sci. Gulf and Caribbean*, v. 14, pp. 126–139.

Schlanger, S. O., 1963, Subsurface geology of Eniwetok Atoll, *U.S. Geol. Survey Prof. Paper 260–BB*, pp. 991–1066.

Shinn, E. A., Ginsburg, R. N., and Lloyd, R. M., 1965, Recent supratidal dolomite from Andros Island, Bahamas, *Soc. Econ. Paleont. and Mineral. Spec. Publ. 13*, pp. 112–123.

Skinner, H. C. W., 1963, Precipitation of calcian dolomites and magnesian calcites in the southeast of South Australia, *Am. Jour. Sci.*, v. 261, pp. 449–472.

——, Skinner, B. J., and Rubin, M., 1963, Age and accumulation rate of dolomite bearing carbonate sediments in South Australia, *Science*, v. 139, pp. 335–336.

Taft, W. H., 1967, Physical chemistry of carbonates, in Chilingar, G. V., Bissell, H. J., and Fairbridge, R. W. (eds.), *Carbonate rocks*, Part B, Elsevier, Amsterdam, pp. 151–168,

——, and Harbaugh, J. W., 1964, Modern carbonate sediments of southern Florida, Bahamas, and Espíritu Santo Island, Baja California, *Stanford Univ. Publ. Geol. Sci.*, v. 8, pp. 133–151.

von der Borch, C. C., 1965a, The distribution and preliminary geochemistry of modern carbonate sediments of the Coorong area, South Australia, *Geochim. et Cosmochim. Acta*, v. 29, pp. 781–799.

——, 1965b, Source of ions for Coorong dolomite formation, *Am. Jour. Sci.*, v. 263, pp. 681–688.

Wells, A. J., 1962, Recent dolomite in the Persian Gulf, *Nature*, v. 194, pp. 274–275.

9
Formation and Alteration of Silica and Clay Minerals

Silicate minerals formed at low temperatures, either by weathering or diagenesis, can be divided into three groups: the silica minerals (quartz and opaline silica), the clay minerals, and other silicates, including zeolites, authigenic feldspars, etc. Discussion is confined in this chapter to the first two groups, which quantitatively far outweigh the third in abundance in sedimentary rocks. Although quartz is mainly detrital, having been derived from the physical disaggregation of crystalline rocks, a considerable proportion is also formed in sediments during diagenesis as chert and silica cement. By contrast, clay minerals owe their origin almost entirely to weathering and/or diagenesis. The purpose of this chapter is to discuss some of the more important chemical processes whereby quartz (plus its common predecessor, opaline silica) and clay minerals are formed during the sedimentary cycle. Topics include the solubility and formation of opaline silica, the origin of chert, the formation of clays during weathering, initial reaction between clays and dissolved species upon entering sea water (halmyrolysis), and clay-mineral diagenesis.

DIAGENESIS OF SILICA

SILICA MINERALS

The silica minerals which form authigenically include quartz and opaline silica.   Quartz, SiO_2, can form as euhedral overgrowths on detrital grains, as a microcrystalline cement in sandstones and shales (see Chapter 6), and as chert.   The quartz in chert occurs as very fine fibrous crystals and is commonly referred to as chalcedonic quartz.   Opaline silica occurs mainly as skeletal components of planktonic organisms and sponges.   It can also form as an inorganic precipitate.   According to Frondel (1962) its structure consists of an aggregate of submicroscopic crystallites of disordered cristobalite plus intercrystalline water.   Disordering is due to this fine crystal size plus the substitution of OH^- for O^{--}.   Water content ranges from 3 to 10 percent by weight.

CHEMISTRY OF DISSOLVED AND OPALINE SILICA

Dissolved silica at pH values less than 9 is present in solution only as monomeric silicic acid, H_4SiO_4 (see Siever, 1971).   At values above 9, H_4SiO_4 dissociates to $H_3SiO_4^-$ and $H_2SiO_4^{--}$.   Since the pH of the ocean and most other natural waters is less than 9, the species involved in the precipitation and dissolution of opaline silica and quartz is usually H_4SiO_4.   Only in rare alkaline lakes need the other silica species be considered.   The concentration of H_4SiO_4 in sea water is variable and ranges from less than 10^{-6} to about 2×10^{-4} mole per liter.   Most of the ocean below a few hundred meters is close to the average value of 10^{-4} m (Turekian, 1968).   Since H_4SiO_4 is a neutral species, its activity coefficient should be close to one at ionic strengths encountered in fresh waters and sea water.   The concentration in rivers and nonalkaline lakes ranges from about 3×10^{-5} to 10^{-3} m (Siever, 1971).   The average value for rivers is 2.2×10^{-4} m (Livingstone, 1963).

The solubility of amorphous silica in the form of synthetic silica gel, colloid, and silica glass has been measured at various temperatures and pH values by Krauskopf (1956).   Equilibrium was approached by dissolution or polymerization from supersaturated solution to form a colloidal sol.   Results are summarized in Table 9-1.   The range at each temperature represents variations with the form of silica used and differences between values obtained by dissolution and polymerization.   The solubility reaction for the pH range 0 to 9 is:

$$SiO_{2\,\text{amorph}} + 2H_2O \leftrightarrows H_4SiO_{4\,\text{aq}}$$

At 25°C:

$$K = a_{H_4SiO_4} = 2 \times 10^{-3} \tag{1}$$

Table 9-1   Solubility of amorphous silica
in water at pH values less than 9.
(Krauskopf, 1956)

Temperature	ppm SiO_2	$m_{H_4SiO_4} \times 10^3$
0°C	60–80	1.0–1.3
25°C	100–140	1.7–2.3
90°C	300–380	5.0–6.3

From Table 9-1 and the measured values reported earlier, Fig. 9-1 is constructed.   It can be seen that average river water and sea water are undersaturated with respect to synthetic amorphous silica.   It is assumed that natural opaline silica, which dissolves too slowly to measure its solubility directly, has a solubility similar to that of the synthetic materials.   (Analyses of pore water of siliceous sediments discussed below bear this out.)   Therefore, rivers and the oceans should be undersaturated with respect to natural opaline silica.

Krauskopf (1956) also studied the rate of dissolution, polymerization, and precipitation (flocculation) of silica gel.   In distilled water, dissolution took several weeks to reach its saturation value and polymerization took several months.   Once a colloidal sol was formed by polymerization it was found to be very stable.   Flocculation of the sol could be accomplished by using a very high initial supersaturation, which resulted in a high sol concentration before precipitation, or by using flocculating electrolytes and oppositely charged sols.   Flocculation and dissolution rates are also pH dependent (Siever, 1971).   Flocculation is decelerated at low or high pH, with the maximum rates occurring in the range pH 4 to 7.   Dissolution may take several months below pH 5, whereas at pH 8 a silica gel may dissolve to saturation overnight.   Because sea water contains abundant electrolyte, and has a pH in the range 7.5 to 8.3, dissolution and precipitation of amorphous silica in the oceans should be rapid.

According to Carman's theory (Iler, 1955) the polymerization of dissolved silica involves the splitting off of water as a result of forming siloxane bonds. The reaction can be represented as

$$2\left(\begin{array}{c} OH \\ | \\ OH-Si-OH \\ | \\ OH \end{array}\right) \rightarrow \left(\begin{array}{c} OH \quad OH \\ | \quad\quad | \\ OH-Si-O-Si-OH \\ | \quad\quad | \\ OH \quad OH \end{array}\right) + H_2O$$

Monosilicic acid   Disilicic acid

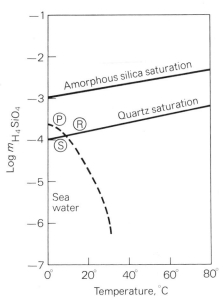

Fig. 9-1 Concentration (molality) of H_4SiO_4 at various temperatures for equilibrium with quartz and amorphous silica and measured values for average river water (R), average sea water (S), and pore waters of deep sea sediments (P). Ranges of values for sea water are enclosed by the dashed line. (*Data from Krauskopf, 1956; Siever, 1962; Sverdrup et al., 1942; Turekian, 1968; Livingstone, 1963; and Siever et al., 1965.*)

The siloxane bond is represented by the –Si–O–Si– portion. Continued siloxane bond formation in three dimensions results in the formation of a silica sol particle. The bonding geometry is very similar to that in the mineral, cristobalite. At the surface of each sol particle Si atoms are coordinated by acidic OH groups, making the sol a polyacid and giving it its characteristic negative charge. The isoelectric point is very low, around pH 1.8 (Parks, 1965). Gel formation occurs when the sol particles coalesce and establish a limited number of siloxane cross-linkages. The resulting structure is highly porous and contains abundant water and impurity cations which were originally positive counterions in the electrical double layers of the sol particles. During aging further cross-linkage formation occurs and part of the water and cations is expelled. Eventually the gel hardens enclosing some water, and the resulting product is opaline silica.

The solubility of amorphous silica in sea water was found by Krauskopf (1956) to be the same as in distilled water. However, the value in sea water is metastable since Wollast et al. (1968) found that Mg^{++} in sea water will form magnesium silicate at concentrations of dissolved silica greater than $4 \times 10^{-4} m$ at 25°C. Precipitation of silica by cations other than Mg^{++} was not observed.

CHERT FORMATION

Chert is a sedimentary rock made up of chemically precipitated, fibrous microcrystals of chalcedonic quartz. It must form during later diagenesis because direct precipitation of chert or any other form of quartz has not been demonstrated in recent sediments.   That some natural waters should precipitate quartz is shown by its solubility.   At 25°C the equilibrium constant for the reaction

$$SiO_{2\,quartz} + 2H_2O \leftrightharpoons H_4SiO_{4\,aq}$$

is

$$K = a_{H_4SiO_4} = 1.7 \times 10^{-4} \tag{2}$$

This value is based on extrapolations from measurements at higher temperatures (Siever, 1962).   Dissolution rate at 25°C, like precipitation rate, is negligible, and grinding does not promote dissolution of quartz as such but instead converts the surfaces of the quartz grains to amorphous silica so that an anomalously high solubility results (Morey et al., 1962).   Since the average value of $a_{H_4SiO_4}$ for river water is 2.2×10^{-4}, many rivers are supersaturated with respect to quartz (see Fig. 9-1).   Most pore waters of modern sediments are also supersaturated with respect to quartz (Siever et al., 1965, and Fig. 9-1), yet there is no evidence for quartz formation in these sediments.

Bedded cherts must have been laid down originally as some form of silica. Since quartz precipitation is unlikely, the originally deposited material must have been amorphous or opaline silica.   Two possibilities exist as to the source of this silica.   One is that it was precipitated inorganically from supersaturated solution, and the other is that it was supplied as the skeletal remains of silica-secreting organisms.   For normal sea water, inorganic precipitation is impossible since it is undersaturated with amorphous silica.   Thus, it seems likely that most cherts were formed by the original accumulation of siliceous organic remains.   (Possible inorganic cherts will be discussed a little further on.)

In certain areas of the ocean, opaline silica skeletal debris is presently accumulating in sufficiently high concentration that it would form chert if recrystallized to chalcedonic quartz.   These include the Gulf of California (Calvert, 1966); southern California borderland (Emery, 1960); Antarctic regions of the ocean, where the sediments contain high proportions of diatom frustrules; and some equatorial deep-sea localities where radiolarian tests are abundant (Sverdrup et al., 1942).

Interstitial waters of the Gulf of California sediments are essentially saturated with amorphous silica at the temperature of the sediments (Siever et al., 1965).   This results from the dissolution of diatoms to equilibrium and causes a negligible loss of silica because of the semiclosed system nature of modern sediments.   Amorphous silica is much more soluble than quartz, as shown above.   Thus the pore waters are supersaturated with respect to quartz, and chert can form thermodynamically.   The standard free-energy change for the transformation of amorphous silica to quartz is related to the solubility equilibrium constants by

$$\Delta F^\circ = RT \ln \frac{K_{quartz}}{K_{amorph}}$$ (3)

For 25°C:

$$\Delta F^\circ = -1.5 \text{ kcal/mole}$$ (4)

Most ancient bedded cherts provide direct evidence of a biogenic origin in the form of fossil diatoms or radiolaria. For cherts where recrystallization has destroyed any possible fossil evidence, it has been suggested by many writers that the original silica was precipitated inorganically. Since normal sea water is undersaturated, special conditions enabling supersaturation have been postulated. The usual theory is that of a nearby volcanic source of the excess silica. There are two major problems with the inorganic volcanic theory. First, biogenic silica secretion in shallow waters of the present ocean is so active that dissolved silica is removed to extremely low concentration levels. Volcanically derived silica should stimulate the growth of siliceous organisms which may easily remove silica faster than it can build up to saturation. Secondly, alteration of volcanic glass in sea water does not lead to amorphous silica formation but rather to crystallization of montmorillonite and zeolites, as shown by bottom sediments of the Pacific Ocean (Arrhenius, 1963).

The most favorable environment for possible volcanic-inorganic chert formation can be predicted. This is a deep, restricted basin where water circulation at depth is minimal and where submarine volcanism takes place. Silica released to solution from volcanic glass because of a lack of circulation is not carried up to shallow water where it can be precipitated by organisms. This enables a buildup of dissolved silica to saturation in the deep water. Because of a lack of circulation, dissolved O_2 is removed by bacterial activity at depth and is not replenished by downward-flowing currents. Thus, the deep water, besides being high in silica, should be anaerobic and most likely sulfidic as in the Black Sea. The common occurrence of bedded cherts in association with black shales and submarine volcanics (eugeosynclinal assemblage) is a predicted result of the conditions necessary for the formation of inorganic chert. Thus, it is possible that some cherts are inorganic.

There are no modern examples, however, of volcanic-inorganic silica precipitation in sea water. One locality, the Gulf of California, fulfills most of the requirements necessary for volcanic chert formation (abundant nearby volcanism, restricted basins of low O_2 content, deep water), but the silica there is all precipitated by diatoms. In fact Calvert (1966) has shown that sufficient silica to account for all diatom sedimentation is provided by inflow to the Gulf from the open ocean and that a volcanic silica source is unnecessary.

Inorganic silica precipitation from volcanic sources in alkaline lakes has been demonstrated (Eugster and Jones, 1968), but these are small isolated bodies which should be very rare occurrences in the geologic column.   It is possible that in the Precambrian, silica secreting organisms had not yet evolved and that most of the nonfossiliferous Precambrian cherts were formed inorganically.

Chert also occurs abundantly in sedimentary rocks as nodules and fossil replacements.   The source of the silica for such "secondary" chert is not as obvious as in the case of bedded cherts.   In fossiliferous carbonate rocks the source of silica is, in many cases, probably the remains of siliceous sponges. Dissolution of the sponges, local migration, and precipitation of silica would be involved in silicification of carbonate shells or nodule formation.   Large-scale silicification of carbonate rocks poses a greater problem regarding an origin for the silica.   Volcanic material cannot be invoked in the case of shelf limestones because of the usual lack of any associated volcanic rocks.   Large-scale silicification thus requires original sedimentation of a relatively large proportion of siliceous organisms or invasion of the rock by ground waters bringing in silica dissolved from areally or stratigraphically adjacent silicate rocks.

Silicification as a result of weathering has been shown by Altschuler et al. (1963).   They found that the present-day weathering of montmorillonite to kaolinite in central Florida results in the liberation of silica which replaces $CaCO_3$ fossils.   Since biogenic CO_2 is abundant during weathering, silicification of calcium carbonate can be explained by the simultaneous dissolution of $CaCO_3$ by carbonic acid and the precipitation of silica (whose solubility is not affected by pH).   The reason why silica precipitation is localized at the site of carbonate dissolution is not clear.

Chert can also form from the dissolution of detrital quartz.   Peterson and von der Borch (1965) have demonstrated that in some South Australian saline lakes and lagoons very high pH values (>10) occur due to photosynthesis. The alkaline waters dissolve detrital quartz grains and produce silica concentrations in excess of that for saturation at lower, more neutral pH.   Upon burial the silica-rich water becomes neutralized by CO_2 produced during organic decomposition.   This causes the solubility of amorphous silica to decrease, because of the conversion of $H_3SiO_4^-$ to H_4SiO_4, and as a result the waters become supersaturated, and opaline silica is precipitated.   The possible importance of high pH waters in the dissolution and transportation of silica in other geologic environments, as suggested by this study, deserves more attention.

The time scale for the conversion of amorphous silica to chalcedonic quartz is not accurately known, but geologic evidence indicates that it takes no longer than about 10 to 100 million years, since older opaline silica is usually absent.   Partial conversion to quartz has been found in Pleistocene cherts indicating that times for transformation are variable and dependent on local geochemical factors.   Recent studies of buried deep-sea Eocene and Cretaceous cherts (Beall and Fischer, 1969) indicate that considerable dissolution of siliceous organisms has occurred but that the silica is redeposited largely as opaline silica and only partly as chalcedonic quartz.

CLAY MINERALS

The common sedimentary clay minerals can be subdivided into five groups: the kaolinite group, the montmorillonoids or smectites, the illites, the chlorites, and the vermiculites.   Each group may contain a few or a large number of compositional and structural varieties, many of which have separate mineral names.   Detailed discussion of crystal structure and varietal species will not be presented here and the reader is referred to the books by Grim (1968) and Brown (1961) for further information.   The present brief discussion is concerned mainly with identifying the chief compositional and structural differences between each group.   The one feature that the five classes have in common is that they all are layer-type aluminosilicates.   A summary of the following descriptions is given in Fig. 9-2.

Kaolinite is by far the most abundant member of the kaolinite group and can be represented relatively accurately by the formula $Al_2Si_2O_5(OH)_4$.   Its crystal structure consists of layers of aluminum ions octahedrally coordinated by OH^- or O^{--} (octahedral layer) regularly alternating with layers of corner-shared silica tetrahedra (tetrahedral layer).   One octahedral plus one tetrahedral layer strongly bound together constitute the basic unit cell, which is approximately 7 Å thick.   There is little ionic substitution for Al or Si, and as a result interfacial electrical double layers due to substitution (type I of Chapter 2) are unimportant.   Ion-exchange capacity is correspondingly low (1 to 10 meq/100 g) due to the lack of substitution.

The mineral montmorillonite is the most abundant member of the montmorillonoid group.   Its unit cell consists of one aluminous octahedral layer sandwiched between two tetrahedral layers.   There is limited substitution of Al^{3+} for Si^{4+} in the tetrahedral layers and extensive substitution of Mg^{++} (and to a lesser extent Fe^{3+} and Fe^{++}) for Al^{3+} in the octahedral layer.   This gives rise to a large net negative charge.   To balance the negative charge exchangeable cations accumulate in the interlayer positions between each unit cell as well as on the flat surface of each crystal.   The cations constitute the counterions of a type I electrical double layer and the interlayer positions can be considered as a part of the total surface area of a montmorillonite crystal. Because of extensive substitution the ion-exchange capacity is high, ranging from 80 to 140 meq/100 g.   The cations are hydrated and as a result variable quantities of water accumulate in the interlayer positions.   Variable hydration gives rise to an expandable structure and a variable thickness for the unit cell. The property of expandability is the most important criterion for distinguishing

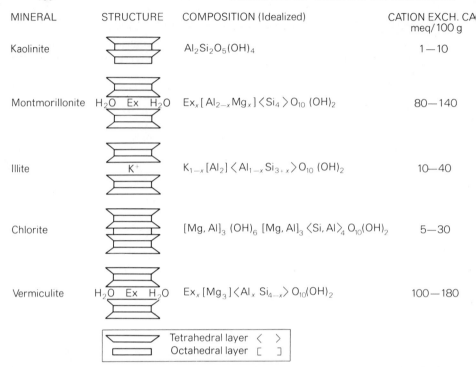

MINERAL	STRUCTURE	COMPOSITION (Idealized)	CATION EXCH. CAP meq/100 g
Kaolinite		$Al_2Si_2O_5(OH)_4$	1—10
Montmorillonite	H_2O Ex H_2O	$Ex_x[Al_{2-x}Mg_x]\langle Si_4 \rangle O_{10}(OH)_2$	80—140
Illite	K^+	$K_{1-x}[Al_2]\langle Al_{1-x}Si_{3+x}\rangle O_{10}(OH)_2$	10—40
Chlorite		$[Mg,Al]_3(OH)_6[Mg,Al]_3\langle Si,Al\rangle_4 O_{10}(OH)_2$	5—30
Vermiculite	H_2O Ex H_2O	$Ex_x[Mg_3]\langle Al_x\ Si_{4-x}\rangle O_{10}(OH)_2$	100—180

Tetrahedral layer $\langle \ \rangle$
Octahedral layer $[\ \ ]$

Fig. 9-2 General characteristics of the principal clay mineral groups. Values for ion-exchange capacities modified from the data of Grim (1968). The symbol Ex stands for hydrated exchangeable cations.

montmorillonite from other clays which do not expand at all or show limited expandability when treated with ion-solvating liquids in the laboratory. An average composition for well-crystallized iron-free montmorillonite (Foster, 1954) is

$$Ex_{0.4}[Al_{1.65}Mg_{0.35}]\langle Al_{0.05}Si_{3.95}\rangle O_{10}(OH)_2 \cdot nH_2O$$

where Ex = exchangeable cations (expressed as univalent ions)

     [ ] = octahedral layer

     $\langle \ \rangle$ = tetrahedral layer

     nH_2O = interlayer water

Numerous other formulas can be found in Ross and Hendricks (1945). Exchangeable cations consist predominantly of the common cations of natural waters, Na^+, Ca^{++}, Mg^{++}, and K^+. The composition above should be considered only as an approximation for any given sedimentary montmorillonite because of considerable variability in composition due to ionic substitution.

Illite can be considered to be a compositional variant of the mineral muscovite. The structure of muscovite consists, like montmorillonite, of an aluminous octahedral layer sandwiched between two tetrahedral layers. In "ideal" muscovite there is no substitution in the octahedral layer and in the tetrahedral layer one out of every four silicons is replaced by an aluminum. The resulting net negative charge is balanced by nonhydrated, nonexchangeable K^+ ions in the interlayer position. This structure gives rise to a 10-Å unit cell and the muscovite composition

$$K[Al_2]\langle AlSi_3 \rangle O_{10}(OH)_2$$

Because the K^+ ions are firmly bound and not hydrated, muscovite is not expandable. Illite is essentially fine-grained muscovite in which there is some octahedral substitution, a lesser amount of aluminum substitution in the tetrahedral layer (and consequently lower K^+ in the interlayer), and some replacement of K^+ by exchangeable cations. An average composition given by Foster (1954) is

$$Ex_{0.2}K_{0.5}[Al_{1.8}Mg_{0.2}]\langle Al_{0.5}Si_{3.5} \rangle O_{10}(OH)_2$$

Due to the presence of a limited number of interlayer exchange sites, the cation exchange capacity of illite is intermediate between that of kaolinite and montmorillonite.

Chlorite is an aluminosilicate of magnesium and iron. The 14-Å unit cell consists of a 2-tetrahedral-plus-1-octahedral "sandwich" combined with an additional octahedral layer which is often called the brucite layer. The principal ion of both octahedral layers is magnesium. Substitution of Al^{3+} for Si^{4+} in the tetrahedral layers is balanced by substitution of Al^{3+} for Mg^{++} in the octahedral layers. Thus there are no interlayer ions. The composition, which is highly variable, can be represented as

$$[Mg,Al]_3\langle Si,Al \rangle_4 O_{10}(OH)_2 \cdot [Mg,Al]_3(OH)_6$$

Ferrous iron substituting for Mg^{++} is almost always present and is a characteristic feature of chlorite. Because of an absence of interlayer hydrated cations the structure is not expandable. The cation-exchange capacity is generally low although in fine-grain sizes it can increase due to a poorly developed brucite layer.

Vermiculite is similar to montmorillonite except that the net negative charge is due mainly to substitution of Al for Si in the tetrahedral layer. The octahedral layer can be either predominantly magnesian (trioctahedral vermiculite) or aluminous (dioctahedral vermiculite) (Jackson, 1964). A generalized composition for trioctahedral vermiculite is

$$Ex_x[Mg_3]\langle Al_xSi_{4-x} \rangle O_{10}(OH)_2 \cdot nH_2O$$

Although the degree of hydration and expandability is less than that for montmorillonite, the cation-exchange capacity is generally higher ranging from 100 to 180 meq/100 g. This is due to extensive substitution of Al for Si (x can be as high as 0.7) in the tetrahedral layer.

Intimately interstratified mixtures (on the unit-cell level) between two or more of the above minerals commonly occur in sediments. The interstratifications can be regularly alternating or random. The most common examples are random-mixed-layer montmorillonite–illite and random or regularly alternating-mixed-layer vermiculite–chlorite. Much material called illite may actually be muscovite with a limited amount of interlayered montmorillonite.

WEATHERING AND CLAY MINERAL FORMATION

During weathering, clay minerals form as a result of the chemical decomposition of silicate rocks by soil waters. The nature of the clay formed depends primarily upon three chemical factors. They are the mineral composition of the rock, the chemical composition of the water, and the rate of passage of water through the rock. In the present section these factors will be discussed and related to the more usually cited weathering controls: climate, rock type, relief, and vegetation. Since chemical weathering is a complex subject which has received a great deal of attention from many different scientific approaches, the following discussion should be considered as only a brief outline relating weathering to the formation of clay minerals. For more details the reader is referred to books devoted to various aspects of weathering, such as those by Bear (1964), Keller (1957), Rich and Kunze (1964), Marshall (1964), and Kononova (1966).

MINERALOGY OF ORIGINAL ROCK

Different silicate minerals weather at different rates. This was pointed out most eloquently by Goldich (1938), who showed that a weatherability series for the common igneous minerals could be constructed (see Fig. 9-3) which resembles very closely the Bowen reaction series of igneous petrology. Olivine and calcic plagioclase are most easily decomposed, followed in decreasing order of weatherability by pyroxenes, amphiboles, and intermediate plagioclase; sodic plagioclase and biotite; potassium feldspar; muscovite; and quartz. To this series should be added volcanic glass, which weathers very rapidly, probably more rapidly than any of the above minerals. Although the Goldich series is an overall guide to weathering, it should be kept in mind that specific exceptions occur due to differences in grain size and other factors. For all practical purposes quartz can be considered to be inert during weathering and it will be neglected in further discussions. In addition, the relatively nonaluminous ferromagnesian minerals (olivine and most pyroxenes and

Fig. 9-3   Weathering stability series for the common igneous minerals. Resistance to chemical weathering increases downward in the list. This is similar to the Bowen reaction series of igneous petrology. The similarity probably arises from the fact that the less stable minerals are the higher temperature minerals, which, at earth surface conditions, are further removed from their original environment of formation. (*After Goldich, 1938.*)

amphiboles) will be only briefly mentioned since clay minerals are highly aluminous and therefore require aluminous precursors. This then leaves plagioclase, K feldspar, biotite, and muscovite. Of the four, plagioclase, because of its high reactivity and abundance, appears to be the principal source of the cations and silica found in most igneous ground waters (Garrels, 1967).

Kaolinite and montmorillonite are clays whose composition or structures are completely different from those of the common igneous minerals. Therefore, they must form by crystal growth from the weathering decomposition products of the primary minerals. Under conditions of limited weathering, the source of the decomposition products is most probably plagioclase, although in the case of montmorillonite volcanic glass is also an important source. In the absence of volcanic glass, ferromagnesian minerals are required for montmorillonite because it invariably contains appreciable magnesium in the octahedral layer of its structure. Under conditions of intense weathering and decomposition, muscovite and K feldspar, as well as plagioclase, may provide the alumina and silica necessary for forming kaolinite and montmorillonite.

Vermiculite forms primarily from the partial weathering of muscovite and biotite (Jackson, 1964). This appears to be reasonable because of the similarity in structure between vermiculite and the micas. They are all layer-type aluminosilicates containing appreciable substitution of Al for Si in the tetrahedral layer. Jackson (1964) describes various schemes for the weathering of micas to illite and vermiculite. In the case of vermiculite, interlayer K^+ in muscovite and biotite is replaced by hydrated cations. This gives rise to its characteristically high cation-exchange capacity.

Illite found in soils is probably to a great extent simply inherited from pre-existing sedimentary illite of diagenetic origin, or from the partial weathering of igneous and metamorphic muscovite. Illites in the clay fraction of modern sediments of the Atlantic Ocean (Hurley et al., 1963; Dasch et al., 1966; Dasch, 1969) have radiometric "ages" similar to the rocks from which they are derived. This militates against an origin by crystal growth from solution during weathering. Some illite is definitely formed by the partial weathering of muscovite (e.g., Weaver, 1958). Illite is generally more siliceous and less potassic than high-temperature muscovite. A reasonable mechanism for the formation of illite (Jackson, 1964) entails the partial replacement of muscovite by montmorillonite layers on an atomic scale, giving rise to a higher silica content, partial lattice expandability, and lowered potassium content. This is in keeping with the idea mentioned earlier that many illites are really mixed-layer mica-montmorillonites. However, an important property of expandable illites derived from the weathering of muscovite, as discussed by Weaver (1958), is the fact that such "degraded" illites will become nonexpandable when placed in a potassium-rich environment such as sea water. This is not true of ideal montmorillonite.

Most sedimentary chlorite is probably derived from pre-existing chlorite, although limited amounts may form from other layer-type silicates during weathering (Jackson, 1964). The predominance of chlorite over kaolinite as a clay component of sediments at high latitudes (Biscaye, 1965) where chemical weathering is relatively weak suggests that it is formed by the disaggregation of pre-existing chlorite.

WATER COMPOSITION AND FLOW RATE

The formation of clays by weathering is the result of the reaction of hydrogen ion in soil waters with primary silicate minerals. The source of H^+ is mainly carbonic acid formed by the bacterial decay of soil organic matter. Other sources of hydrogen ion (Keller, 1957) are organic acids, exchangeable H^+ on acid clays and plant rootlets, and in the absence of vegetation, water itself. In the latter case the weathering reaction is known as hydrolysis.

The mechanism of the reaction of hydrogen ion with K feldspar has been studied experimentally by Wollast (1967). Wollast suspended samples of ground orthoclase in a series of buffered solutions of different pH (4 to 10) and measured the rate of release of dissolved silica and alumina to solution. The initial reaction taking place over a time scale of 30 minutes was simply the irreversible ion exchange of K^+ on the surface of feldspar for H^+ in solution:

$$H_{aq}^+ + K \, spar_{surface} \rightarrow H \, spar_{surface} + K_{aq}^+$$

The rate of this reaction was found to be a direct function of pH as expected.   However, hydrogen feldspar is very unstable and breaks down rapidly, releasing dissolved silica and alumina to solution.   At the higher pH values used (>5) dissolved aluminum cannot build up in solution and is precipitated as $Al(OH)_3$.   Thus, Wollast found that the concentration of dissolved aluminum rapidly leveled off at the value expected for equilibrium with freshly precipitated (amorphous) $Al(OH)_3$.

The rate of increase of H_4SiO_4 during Wollast's experiments is shown in Fig. 9-4.   Note that the concentration of H_4SiO_4 begins to level off at values far less than that expected for equilibrium with amorphous silica $(2 \times 10^{-3} \, m)$.   This indicates that the silica is probably removed from solution by some other chemical reaction.   To fit his experimental data, Wollast adopted the following model (Fig. 9-5): the $Al(OH)_3$ which is rapidly precipitated forms a protective surface layer on the feldspar.   The rate of dissolution of feldspar is assumed to be diffusion-controlled but retarded by diffusion through the surface layer.   Diffusion-controlled dissolution at constant surface area into a closed system, the situation present in Wollast's experiments, is described by Eq. (98) of Chapter 2:

$$\frac{dC}{dt}_{\text{diss}} = \frac{D\bar{A}}{l}(C_S - C) \tag{5}$$

where C = concentration of H_4SiO_4

$\bar{A}$ = area of feldspar surface per unit volume of solution

l = thickness of $Al(OH)_3$ layer

D = diffusion coefficient in the $Al(OH)_3$ layer

C_S = solubility of amorphous silica $(2 \times 10^{-3} \, m)$

t = time

Now, the silica in solution is that which was originally contained within the aluminous surface layer so that

$$C = C_F \bar{A} l \tag{6}$$

where C_F = moles of SiO_2 per cm^3 in feldspar.   Thus, substituting into (5) for l

$$\frac{dC}{dt}_{\text{diss}} = DC_F \bar{A}^2 \left(\frac{C_S - C}{C} \right) \tag{7}$$

This expression describes the initial portion of the curve in Fig. 9-4 where silica concentrations are less than 8×10^{-5} mole/liter.

Flattening of the curve is attributed by Wollast to the reaction of dissolved H_4SiO_4 with $Al(OH)_3$ at the outer surface of the protective layer,

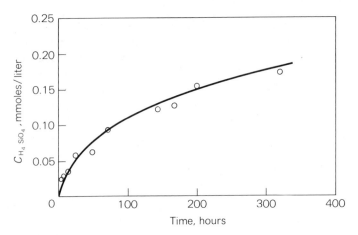

Fig. 9-4 Plot of concentration of H_4SiO_4 versus time for the experimental "weathering" of potassium feldspar. pH = 6 and 5 percent feldspar is in suspension. The curve represents Eq. (9) fitted to the points. (*After Wollast, 1967.*)

resulting in the formation of an amorphous aluminosilicate similar to kaolinite. For this precipitation reaction he assumed a rate law of the type

$$\frac{dC}{dt_{pptn}} = -k(C - C_E) \tag{8}$$

where C_E = equilibrium concentration for the reaction

$$2Al(OH)_{3\ amorph} + 2H_4SiO_{4\ aq} \leftrightharpoons Al_2Si_2O_5(OH)_{4\ amorph} + 5H_2O$$

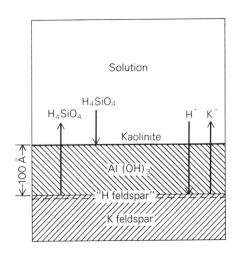

Fig. 9-5 Diagrammatic representation of the experimental feldspar weathering model of Wollast (1967).

For crystalline $Al(OH)_3$ (gibbsite) and kaolinite the value of C_E calculated from the data of Appendix I is 1.1×10^{-5} mole/liter. Combining Eqs. (7) and (8), Wollast obtained the overall equation for C as a function of time:

$$\frac{dC}{dt}_{\text{total}} = -k(C - C_E) + DC_F \bar{A}^2 \left(\frac{C_S - C}{C}\right) \tag{9}$$

By a combination of mathematical simplification and graphical integration, curves of C versus t were obtained which could be fitted well to the experimental points. This is shown in Fig. 9-4. By curve-fitting the values for C_E and D were obtained:

$$C_E = 8 \times 10^{-5} \text{ mole/liter}$$

$$D \sim 10^{-14} \text{ cm}^2/\text{sec}$$

The value for C_E is not unreasonable for equilibrium between x-ray amorphous equivalents of gibbsite and kaolinite. The value for D shows that diffusion is highly retarded by the $Al(OH)_3$ layer because values for ordinary water are on the order of 10^{-5} cm^2/sec. Further tests of Eq. (9) by changing the ratio of ground feldspar to water showed that the predicted dependency upon $\bar{A}$ is correct.

Wollast's experiments were carried out in closed vessels, and as a result, H_4SiO_4 could build up to concentrations sufficiently high to bring about the formation of kaolinite. If an open system had been employed where fresh silica-free water was supplied fast enough so that C was always less than C_E, then continued dissolution of the feldspar would result in the formation of only $Al(OH)_3$. *This points to the importance of water flow as a controlling factor in weathering.* At high rates of flow relative to dissolution, the weathering of feldspar (here using albite) can be represented as

$$H^+ + NaAlSi_3O_{8\text{ albite}} + 7H_2O \rightarrow Al(OH)_{3\text{ gibbsite}} + Na^+ + 3H_4SiO_{4\text{ aq}}$$

At lower flow rates the reaction is

$$2H^+ + 2NaAlSi_3O_{8\text{ albite}} + 9H_2O \rightarrow$$
$$Al_2Si_2O_5(OH)_{4\text{ kaolinite}} + 2Na^+ + 4H_4SiO_{4\text{ aq}}$$

At flow rates approaching stagnancy a further slow reaction, that of cations with $Al(OH)_3$ and silica, can take place resulting in the formation of montmorillonite. Assuming that dissolved Mg^{++} is present, the overall weathering reaction of albite to montmorillonite of average composition would be

$$Mg^{++} + 3NaAlSi_3O_{8\text{ albite}} + 4H_2O \rightarrow$$
$$2Na_{0.5}Al_{1.5}Mg_{0.5}Si_4O_{10}(OH)_{2\text{ montmorillonite}} + 2Na^+ + H_4SiO_{4\text{ aq}}$$

   Natural observations of the occurrences of kaolinite, montmorillonite, and aluminum hydroxide minerals (bauxite) are in agreement with the above predictions.   Bauxite is found only in regions of intense rainfall combined with relatively rugged relief.   Intense rainfall produces a rapid bathing of the rock with dilute water, and rugged relief promotes good drainage and rapid flow through the rock.   These conditions are characteristic of tropical and sub-tropical upland regions with seasonally distributed high rainfall.   An example is Jamaica where bauxite is mined for aluminum.   The more common weathering product of feldspar is kaolinite, especially in tropical regions.   Montmorillonite is associated characteristically with relatively dry climates where flushing rates are low and only the more reactive rock components, especially volcanic glass, are decomposed.   A typical example of montmorillonite formation is the Great Plains region of the United States.   The effect of varying rainfall upon weathering of the same rock is well exemplified by the island of Hawaii (see Fig. 9-6).   All three minerals, bauxite, kaolinite, and montmorillonite, form from an original basalt, but their occurrence depends strongly upon the local mean annual rainfall.

   As mentioned above, relief is also an important control on the rate of flushing of the original rock by water.   Mohr and van Baren (1954) have cited examples of kaolinite and montmorillonite, each forming from the same rock, in tropical regions of high rainfall.   The only difference was that montmorillonite was found in swampy lowlands, whereas kaolinite occurred in nearby well-drained uplands.   Apparently in the swampy lowlands, low

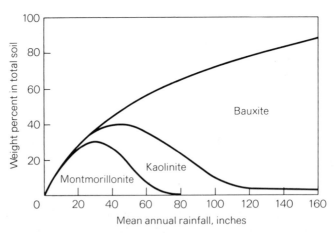

Fig. 9-6   Plots of weight percent in total soil of bauxite, kaolinite, and montmorillonite group minerals versus mean annual rainfall, island of Hawaii.   Vertical distances between successive curves represent concentrations of each mineral group.   Note the increase in total weathering products (top curve) with increase in rainfall. (*After Sherman, 1952.*)

ground-water flow allows sufficient time for the reaction of dissolved cations with alumina and silica to form montmorillonite.

Since the principal source of H^+ in weathering solutions is CO_2 (and H_2CO_3) formed from the bacterial decay of soil organic matter, weathering reactions can be written in terms of CO_2 or H_2CO_3 rather than H^+ (Garrels, 1967). For example, for albite weathering to kaolinite

$$2H_2CO_3 + 9H_2O + 2NaAlSi_3O_8 \rightarrow$$
$$Al_2Si_2O_5(OH)_4 + 4H_4SiO_{4\,aq} + 2Na^+ + 2HCO_3^-$$

This type of reaction explains why the major anion in water from weathered igneous rocks is HCO_3^-. Assuming that bacteriogenic CO_2 reacts with igneous rock, the rate of water flushing relative to rate of rock dissolution can be expressed by the HCO_3^- concentration of the water. Extensive dissolution combined with a slow flow rate should cause a large buildup of HCO_3^- in the water. Thus, more readily weatherable rocks should be associated with higher HCO_3^- concentrations for a constant flow rate. This correlation is demonstrated by the association of high HCO_3^- concentrations in waters draining fine-grained (and presumably glassy) extrusive rocks (Garrels, 1967). Accompanying increased HCO_3^- concentration should be higher concentrations of cations and silica and, if the system is closed to CO_2, a higher pH due to the neutralization of carbonic acid. Thus, montmorillonite formation should also be correlated with higher HCO_3^- concentrations. This can be seen by the reaction

$$3Al_2Si_2O_5(OH)_{4\,kaolinite} + 2Mg^{++} + 2Na^+ + 6HCO_3^- + 10H_4SiO_{4\,aq} \rightarrow$$
$$4Na_{0.5}Al_{1.5}Mg_{0.5}Si_4O_{10}(OH)_{2\,montmorillonite} + 6H_2CO_3 + 19H_2O$$

Increased cations, silica, and HCO_3^-/H_2CO_3 all drive the reaction to the right. Again Garrels presents data which suggests that montmorillonite saturation is reached only in ground waters of high HCO_3^- content (greater than 100 ppm).

Caution should be applied in using HCO_3^- in ground waters as a measure of the weathering of silicates. This is because small traces of $CaCO_3$ may be the source of much of the bicarbonate. Calcite weathers much more rapidly than silicate minerals by the reaction

$$H_2CO_3 + CaCO_3 \rightarrow Ca^{++} + 2HCO_3^-$$

In order to study igneous-rock weathering by means of ground waters, it must be ascertained that there is no nearby carbonate which could serve as a source of HCO_3^-. Presumably this was done in Garrels' study.

HALMYROLYSIS AND EARLY DIAGENESIS OF CLAY MINERALS

After weathering, clay minerals are delivered by erosion to rivers where little further alteration takes place during transport to the ocean. Upon encountering sea water the clays are suddenly thrust into a chemical environment radically different from that during weathering.   The overall composition, ionic strength, and cationic ratios of sea water are considerably different than those for soil water or river water.   As a result chemical reactions take place in an attempt to reach equilibrium.   Some reactions are so rapid that they occur before the clays are buried in bottom sediments.   The term *halmyrolysis* will be applied to these rapid reactions.   Those taking place in the top few meters of sediment are covered by the term early diagenesis.

One of the first halmyrolytic reactions to occur when clays enter sea water is cation exchange.   The ionic composition, in terms of activities, of average river water and sea water is listed in Table 9-2.   Also included are ratios between the monovalent and divalent cations.   Since cation exchange is a rapid process, it would be expected from Table 9-2 that clay minerals, shortly after passing from river water into sea water, would take up magnesium and sodium in exchange for calcium and potassium.   Experimental studies agree with this prediction.   Potts (quoted in Keller, 1963) placed montmorillonitic clay from the Missouri River in sea water in the laboratory and determined subsequent changes in the exchangeable cations on the clay.   His results are summarized in Table 9-3.   Note the increases in Mg^{++}/Ca^{++} and Na^+/K^+ on the clay.   Carroll and Starkey (1960) performed the same experiment using pure illite, montmorillonite, kaolinite, and mixed-layer clay and for all minerals also found that, over a period of days, considerable exchange of Mg^{++} for Ca^{++} occurred in sea water.

Table 9-2   Activities and activity ratios of the principal dissolved species in average river water and sea water.   Data are taken from Chapter 3

Species	*Activity* $\times 10^3$	
	River water	*Sea water*
HCO_3^-	0.89	0.92
SO_4^{--}	0.09	1.94
Cl^-	0.21	346
Ca^{++}	0.30	2.39
Mg^{++}	0.14	13.5
Na^+	0.26	332
K^+	0.06	6.2
Ratio		
Mg^{++}/Ca^{++}	0.45	5.7
Na^+/K^+	4.5	53.5

Table 9-3   Exchangeable cations on Missouri River Clay
after exposure to sea water for the times shown.
[Data from Keller (1963)]

	meq/100 *g dry clay*		
Cation	*Original clay*	*36 hours*	*86 hours*
Ca^{++}	60.7	38.3	17.3
Mg^{++}	20.1	29.7	39.3
Na^+	1.7	1.8	3.4
K^+	1.4	1.4	2.0
Total cations	83.9	71.2	62.0
Molar ratio			
Mg^{++}/Ca^{++}	0.33	0.73	2.3
Na^+/K^+	1.2	1.3	1.7

Russell (1970) repeated Potts' experiment on another river, the Rio
Ameca, Mexico, but did so on a much more comprehensive basis.   Nonexchangeable as well as exchangeable cations were determined.   (Exchangeable ions are defined as those that are replaceable in the laboratory by NH_4^+
at pH 7.)   Also, Russell determined the cation composition, both exchangeable and nonexchangeable, of clays in surface samples of marine sediments
taken within a short distance of the river mouth.   The marine clays were found
to be mineralogically identical to those carried by the river and therefore presumably derived from it.   The principal clay minerals were montmorillonite
(60 percent) and kaolinite (35 percent).   The marine clays are important in
that they provide a longer term experiment on halmyrolysis than can be
achieved in the laboratory.

Some of Russell's results are summarized in Table 9-4.   Note that
Potts' observations are confirmed for another montmorillonitic river clay.
After two weeks' exposure to sea water distinct increases in the ratios
Mg^{++}/Ca^{++} and Na^+/K^+ occurred.   Because chemical analyses were made of
the river water at the time the river clay was taken, the results of Table 9-4 can
be used to test a simple ion-exchange model.   For the exchange reaction

$$Ca_{aq}^{++} + Mg\ clay \leftrightharpoons Mg_{aq}^{++} + Ca\ clay$$

the equilibrium constant is

$$K = \frac{a_{Mg^{++}}\, a_{Ca\ clay}}{a_{Ca^{++}}\, a_{Mg\ clay}} \tag{10}$$

Table 9-4   Exchangeable cations on clays derived
from the Rio Ameca, Mexico, before and after
exposure to seawater in the laboratory.   [Data from
Russell (1970)]

Cation	*meq*/100 *g dry clay*	
	Original clay	*2 weeks in sea water*
Ca^{++}	48	17
Mg^{++}	10	38
Na^+	6	7
K^+	12	4
Total cations	76	66
Molar ratio		
Mg^{++}/Ca^{++}	0.21	2.2
Na^+/K^+	0.50	1.8

If ion exchange is assumed to take place predominantly on one type of exchange
site, Eq. (10) is valid.   For the case of the Rio Ameca clay, because the cation-
exchange capacity of montmorillonite is about ten times as great as that of
kaolinite (see Fig. 9-3), the total exchange capacity can be assumed to be
essentially all due to montmorillonite.*   The exchange sites are the interlayer
positions of montmorillonite which constitute a type I double layer (see
Chapter 2).   If Eq. (10) is valid for Russell's experiment, and if the exchangeable ions can also be described by an ideal solution model, then

$$K = \frac{a_{Mg^{++}} \, X_{Ca \ clay}}{a_{Ca^{++}} \, X_{Mg \ clay}} \tag{11}$$

where X = mole fraction.   Now for the Rio Ameca (Russell, 1970),

$$\frac{a_{Mg^{++}}}{a_{Ca^{++}}} = 0.47$$

Combining this value with the Mg^{++}/Ca^{++} ratio on the original clay given in
Table 9-4,

$$K = 2.2$$

---

* Some of the exchange capacity, however, may also be due to amorphous clay material
whose capacity is not known.

From this value of K, and the Mg^{++}/Ca^{++} ratio for river clay reacted with sea water given in Table 9-4, the value of $a_{Mg^{++}}/a_{Ca^{++}}$ for sea water can be calculated for ion-exchange equilibrium via the above model. Rearranging (11)

$$\left(\frac{a_{Mg^{++}}}{a_{Ca^{++}}}\right)_{SW} = 2.2\left(\frac{X_{Mg}}{X_{Ca}}\right)_{clay\ in\ SW} = 4.9 \tag{12}$$

This value is not far different from that given in Table 9-2 (5.7). This suggests that on two weeks' exposure of montmorillonitic river clay to sea water, cation exchange can be described, as a first approximation, by a simple ideal solution model.

Actually cation exchange on clays is more complicated. Walton (1949) has shown that exchange on clays and synthetic aluminosilicates is well represented by empirical expressions of the type

$$K = \frac{a_{A^+}}{a_{B^+}}\left(\frac{X_{B\ clay}}{X_{A\ clay}}\right)^n \tag{13}$$

with values of n falling generally between 0.8 and 2.0. The above data for the Rio Ameca clay when closely fitted to this expression results in the values $n = 1.06$ and $K = 2.5$. If one type of exchange site is predominant, Eq. (13) can be shown to be a consequence of assuming that the exchangeable ions constitute a symmetrical regular solution (see Chapter 2) on the clay. This has been done by Garrels and Christ (1965, p. 274), who found that for values of X between 0.2 and 0.8, $n = 1 - A/2$ ($A =$ symmetrical regular solution coefficient). Only when $n = 1$ ($A = 0$) is the assumption of an ideal solution strictly valid.

Extended reaction of the Rio Ameca clay with sea water results in a change in its exchange properties. This is shown by the data for the marine clay in Table 9-5. Here the ratio of Mg^{++}/Ca^{++} on the exchangeable sites is 5.2, and the calculated values of n and K are 0.78 and 1.6 respectively. This change is probably caused by a loss of exchangeable sites. Decrease in total cation-exchange capacity during soaking of clays in sea water is a general phenomenon. This is shown by the results of Carroll and Starkey (1960) as well as the data of Potts and Russell listed in Tables 9-3, 9-4, and 9-5. Apparently during reaction of sea water with montmorillonitic river clay, ion fixation takes place, resulting in loss of specific exchange sites with consequent changes in cation-exchange properties.

Table 9-5 shows that over two-thirds of the loss in exchange capacity can be accounted for by the fixation of K^+. This process of irreversible K^+ uptake has been well documented. Weaver (1958) has shown that fully expandable "montmorillonite" derived from the weathering of muscovite, when placed in KOH solution or sea water, readily takes up K^+ and becomes nonexpandable; i.e., it become illitic. Montmorillonite formed by the alteration of volcanic

Table 9-5   Exchangeable (exch.) and fixed (nonexchangeable) cations on clay from the Rio Ameca,
Mexico, and from the surface of marine sediments near
the mouth of the Rio Ameca.   [Data from Russell (1970)]

	meq/100 *g dry clay*			
	River clay		*Marine clay*	
Cation	*Exch.*	*Fixed*	*Exch.*	*Fixed*
Ca^{++}	48	3	6	14
Mg^{++}	10	132	31	120
Na^+	6	13	6	23
K^+	12	19	3	40
Total	76	167	46	197
Total exch. plus fixed	243		243	

glass does not do this.   It thus appears that there are two types of montmorillonite.   The material which reacts with K^+ is considered a potassium-
depleted variety of illite in which a large amount of K^+, during weathering,
has been replaced by exchangeable cations.   To this material the term
degraded illite (dioctahedral vermiculite?) has been applied by other workers.
It is possible that part of the montmorillonite in the river clays used by Potts
and Russell was actually highly weathered, degraded illite which reacted
irreversibly with K^+ in sea water and became normal illite, thus reducing the
cation-exchange capacity.   In Russell's material the degraded illite would have
to exist as interlayers within montmorillonite, rather than as a discrete phase,
because no distinct increase in illite was found during halmyrolysis of the
river clay.   Besides illite, vermiculite formed from biotite weathering also
may pick up K^+ irreversibly from sea water, as shown by Weaver (1958).
Weaver's experimental work is confirmed by the observations of Johns and
Grim (1958).   They have shown that a considerable portion of expandable
illite carried by the Mississippi River picks up potassium irreversibly upon
entry into the Gulf of Mexico.

Table 9-5 also demonstrates increases in fixed, or nonexchangeable, Na^+
and Ca^{++} during halmyrolysis.   Accompanying these increases is a decrease
in fixed magnesium.   Perhaps some of this apparent fixation, including K^+,
is due to slow ion exchange on less readily accessible cation sites. If the
exchange is reversible but slow enough it would not be detected during determination of exchangeable ions by the usual laboratory method which employs
rapid displacement of ions by NH_4^+.   To test this hypothesis, clays which

have been suspended in sea water over long periods should be resuspended in solutions of a greatly different composition and checks made for reversible displacement of the "nonexchangeable" ions.

So far the discussed reactions between clays and sea water have involved only the interlayer ions.   An additional question is whether basic structural changes can take place during halmyrolysis.   The distribution of clay minerals in the oceans suggests that only a minor proportion of clays may undergo complete structural transformation.   Johns and Grim (1958) and Griffin (1962) have shown that the clays in surface sediments of the northern Gulf of Mexico are readily correlated with their river sources.   In addition the montmorillonite-illite ratio of the marine part of the Mississippi Delta is essentially the same as that of suspended material in the river.   Biscaye (1965) in a very comprehensive study has shown that clay minerals in the Atlantic Ocean are almost entirely derived from the surrounding continents.   Montmorillonite is definitely forming over considerable areas of the deep Pacific Ocean bottom, but only from devitrification of volcanic glass and not from other clay minerals (Peterson and Griffin, 1964).

The formation of cationic clays from kaolinite, or cation-free clay in general, during halmyrolysis has been predicted by Mackenzie and Garrels (1966).   Their reasoning is that HCO_3^- delivered by streams to the ocean cannot all be removed by the precipitation of $CaCO_3$:

$$2HCO_3^- + Ca^{++} \rightarrow CaCO_3 + CO_2 + H_2O$$

This is because there are more than two bicarbonate ions for each calcium ion in average river water.   The excess HCO_3^- if not removed would accumulate in the oceans and raise the pH of sea water, over millions of years, to unreasonably high values.   One way to remove HCO_3^- is to react cation-free clay with K^+, Na^+, or Mg^{++} by reactions of the type

$$2K_{aq}^+ + 2HCO_{3\,aq}^- + 3Al_2Si_2O_5(OH)_4 \rightarrow$$
$$2KAl_3Si_3O_{10}(OH)_2 + 5H_2O + 2CO_2$$

This is the reverse of a typical weathering reaction, and besides removing HCO_3^- from sea water, such "reverse weathering" enables the return of CO_2, originally lost during weathering, back to the atmosphere.   The degree of authigenic clay formation necessary to remove all excess incoming HCO_3^- from sea water was calculated by Mackenzie and Garrels (1966) to be about 7 percent of the total clays carried by streams.   Actually this amount should be lower by about half because they incorrectly calculated the amount of calcium, available for HCO_3^- precipitation, which is carried on the exchange sites of clays and released to solution upon contact with sea water.

The results of Russell (1970) discussed earlier have a direct bearing on Garrels and Mackenzie's hypothesis.   The Rio Ameca clay contains abundant cation-free kaolinite. Much of it is of poor crystallinity and fine grain size, and is therefore potentially reactive. If a reverse weathering type of reaction takes place during halmyrolysis, the marine clays derived from the Rio Ameca should show a higher total cation content than the original river clay. As pointed out by Russell, this is not true. The data of Table 9-5 show that the marine and river clays have identical concentrations of total (exchangeable plus fixed) cations. Thus, the Garrels and Mackenzie hypothesis needs another source for verification—or alternative hypotheses to account for the removal of excess HCO_3^- from sea water must be considered.

Both Deffeyes (1965) and Russell (1970) found experimentally that with prolonged soaking montmorillonite will remove Mg^{++} from sea water. Accompanying the loss of Mg^{++} is a decrease in alkalinity. The reaction is highly sensitive to pH and is greatly increased in extent as the pH rises above 8.0. Both authors interpret the results as an uptake of $Mg^{++} + 2OH^-$ in interlayer positions of montmorillonite to produce a chloritelike clay. If true, the reaction would be a mechanism for removing excess alkalinity from the ocean as required by the Garrels and Mackenzie model. The experimental work, unfortunately, does not seem to apply to the natural situation. Russell found no uptake of additional cations, beyond those originally present, in Rio Ameca clay taken from the marine environment. Also, there does not appear to be any evidence for wholesale conversion of montmorillonite to chlorite during halmyrolysis. Perhaps the Mg^{++} plus OH^- uptake is reversible and ions taken up in sea water are released back to solution when the clays are exposed to the lower pH values (7.0 to 7.7) characteristic of marine sediments (Siever et al., 1965).

The clay minerals expected to form in sea water could be predicted if sufficient thermodynamic data were available. Unfortunately at the time of writing of this book, good data were available only for minerals in the system $K_2O-Al_2O_3-SiO_2-H_2O$. This includes kaolinite, muscovite, microcline ($KAlSi_3O_8$), and gibbsite [$Al(OH)_3$]. The relative stability of these phases can be expressed in terms of the parameters $\log a_{K^+}/a_{H^+}$ and $\log a_{H_4SiO_4}$ (Garrels and Christ, 1965). This can be seen by the equilibrium reactions:

$$2Al(OH)_{3\ gibbsite} + 2H_4SiO_{4\ aq} \leftrightharpoons Al_2Si_2O_5(OH)_{4\ kaolinite} + 5H_2O$$

$$K = \frac{1}{a_{H_4SiO_4}^2} = 10^{9.9} \tag{14}$$

$$\log a_{H_4SiO_4} = 10^{-4.95} \tag{15}$$

$$2H_{aq}^+ + 2KAl_3Si_3O_{10}(OH)_{2\ muscovite} + 3H_2O \leftrightharpoons$$
$$3Al_2Si_2O_5(OH)_{4\ kaolinite} + 2K_{aq}^+$$

$$K = \frac{a_{K^+}^2}{a_{H^+}^2} = 10^{9.4} \tag{16}$$

$$\log \frac{a_{K^+}}{a_{H^+}} = 4.7 \tag{17}$$

$$2H_{aq}{}^+ + 2KAlSi_3O_{8 \text{ microcline}} + 9H_2O \leftrightharpoons$$
$$Al_2Si_2O_5(OH)_{4 \text{ kaolinite}} + 4H_4SiO_{4 \text{ aq}} + 2K_{aq}{}^+$$

$$K = \frac{a_{K^+}{}^2}{a_{H^+}{}^2} a_{H_4SiO_4}^4 = 10^{-5.1} \tag{18}$$

$$\log \frac{a_{K^+}}{a_{H^+}} + 2 \log a_{H_4SiO_4} = -2.55 \tag{19}$$

$$2H_{aq}{}^+ + 3KAlSi_3O_{8 \text{ microcline}} + 12H_2O \leftrightharpoons$$
$$KAl_3Si_3O_{10}(OH)_{2 \text{ muscovite}} + 6H_4SiO_{4 \text{ aq}} + 2K_{aq}{}^+$$

$$K = \frac{a_{K^+}{}^2}{a_{H^+}{}^2} a_{H_4SiO_4}^6 = 10^{-12.1} \tag{20}$$

$$\log \frac{a_{K^+}}{a_{H^+}} + 3 \log a_{H_4SiO_4} = -6.05 \tag{21}$$

$$2H_{aq}{}^+ + 2KAl_3Si_3O_{10}(OH)_{2 \text{ muscovite}} + 18H_2O \leftrightharpoons$$
$$6Al(OH)_{3 \text{ gibbsite}} + 6H_4SiO_{4 \text{ aq}} + 2K_{aq}{}^+$$

$$K = \frac{a_{K^+}{}^2}{a_{H^+}{}^2} a_{H_4SiO_4}^6 = 10^{-20.1} \tag{22}$$

$$\log \frac{a_{K^+}}{a_{H^+}} + 3 \log a_{H_4SiO_4} = -10.05 \tag{23}$$

Values of the equilibrium constants K are for 25°C and are derived from the $\Delta F°$ data of Appendix I.

A diagram based on the above equilibrium expressions which shows the fields of stability of the various phases is presented in Fig. 9-7. Also plotted are the saturation curves for amorphous silica and measured values for sea water. The diagram suggests that kaolinite is unstable relative to illite (muscovite) in sea water. Sedimentological observations, however, do not provide evidence for wholesale conversion of kaolinite to illite during early diagenesis. Dasch (1969) has shown that illite in deep-sea sediment, even when very fine grained (<0.1 micron), has a strontium isotope ratio characteristic of detrital illite and not that expected for isotopic exchange and equilibration with sea water. If some illite can form in sea water, it should show evidence for isotopic equilibration in very fine particles.

Because of a lack of reliable thermodynamic data, and because of its extreme compositional variability, montmorillonite is not plotted on a diagram similar to Fig. 9-7 to illustrate its stability in sea water. It is probably more

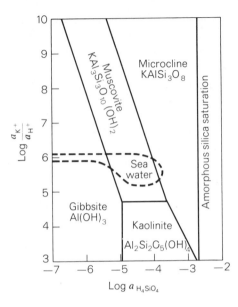

Fig. 9-7 Stability field diagram for common minerals in the system $K_2O-Al_2O_3-SiO_2-H_2O$. $T = 25°C$, $P_{total} = 1$ atm. The area enclosed by dashed lines includes values measured in surface sea water (long narrow portion to the left) and deep sea water (enlarged area to the right). (*Modified from Garrels and Christ, 1965.*)

stable than kaolinite since it forms during the submarine alteration (or weathering) of volcanic glass (Peterson and Griffin, 1964). Sea water is high in cations and by analogy with terrestrial weathering, montmorillonite rather than kaolinite would be expected to be the chief clay formed by submarine weathering. On the other hand, the experiment of Deffeyes and Russell discussed earlier suggests that some sort of chloritelike derivative of montmorillonite may be more stable in sea water than montmorillonite itself. The experimental work of Whitehouse and McCarter (1958) suggests that montmorillonite is unstable relative to illite and/or chlorite in sea water. Whitehouse and McCarter suspended samples of montmorillonite in artificial sea water for periods ranging up to five years and found that with time the product could be separated, by differential settling rates, into original material plus small proportions of newly formed illitic and chloritic phases. A problem with their results, however, is that the starting material was not also fractionated to check for the possible presence of minor amounts of degraded illite which, by simple exchange of interlayer cations, might be reconstituted to true illite in sea water. If Whitehouse and McCarter's conclusions are correct, one should observe wholesale conversion of montmorillonite to illite (plus chlorite) during halmyrolysis and early diagenesis. As mentioned earlier, the evidence for any such extensive transformation is lacking. A possible explanation is that dissolved organics, especially in the pore water of surficial sediments, may inhibit transformation. This was shown in experiments conducted by Whitehouse and McCarter, where raffinose and fucoidin added to the artificial sea water caused a severe inhibition of chlorite and illite formation. The explanation is that the dissolved organics block interlayer sites otherwise available for uptake of cations. Certainly much more work is needed in order to clear up the many apparent contradictions between experimental studies and natural observations of clay minerals in sea water.

Up to this point discussion has been directed almost entirely toward the halmyrolysis of clay minerals rather than early diagenesis. Early diagenesis, however, provides little evidence for the structural conversion of detrital clays to authigenic clays. No consistent change in clay mineralogy with depth, attributable solely to diagenesis, is found in cores of recent sediments taken from different localities. Those changes observed can almost always be attributed to changes in the detrital clay content at the time of deposition. These changes are often caused by differential sedimentation of the detrital load (Parham, 1966). Analyses of pore waters for dissolved cations and silica in a variety of clay-rich modern marine sediments (e.g., Siever et al., 1965) also provide little evidence for early diagenetic clay formation. Increased concentrations of H_4SiO_4 found in most pore waters can be readily traced to the dissolution of opaline silica contained in diatom frustrules, radiolarian tests, etc., and not to clay mineral-forming reactions. Apparently early diagenesis is a relatively uninteresting period during the geological history of clay minerals. Most reactions take place during weathering, halmyrolysis, or later diagenesis.

LATER DIAGENESIS OF CLAY MINERALS

The amount of time represented by the top few meters of unlithified sediment (i.e., early diagenesis) represents at the very most a few million years. Temperatures rarely exceed 30°C. At such low temperatures, recrystallization of one clay mineral to another, although favored thermodynamically, may not take place during early diagenesis because of the presence of high activation energy barriers. Higher temperatures and/or more time are needed. If major diagenetic changes in clay mineralogy actually occur, evidence for them must be sought through the study of deeply buried sediments and/or ancient sedimentary rocks.

Studies of Paleozoic and Mesozoic clay-bearing sedimentary rocks provide considerable evidence that diagenetic changes do occur during later diagenesis. Weaver (1967), after studying a large number of Phanerozoic shales, has concluded that clays older than the Upper Paleozoic consist for the most part of only illite and chlorite. Younger rocks exhibit a more varied clay mineralogy including montmorillonite, mixed-layer clays, and kaolinite, as well as illite and chlorite. Although Weaver chose to interpret his observations as indicating a change in the type and degree of weathering with time, the data can also be interpreted as showing a progressive diagenetic conversion to illite and chlorite over long geologic times.

The presence of low-temperature illite in Paleozoic shales has been conclusively demonstrated by Velde and Hower (1963). They have found that the finer illite fraction consists predominantly of the 1Md polymorphic modification of mica* which does not form above temperatures of about 105°C (Velde, 1965) and is therefore not found in igneous or metamorphic rocks. The illite must be formed by weathering or diagenesis. Suggestive evidence that some Paleozoic illite is diagenetic is provided by K-Ar dating (Hower et al., 1963). In an Upper Ordovician shale, the low-temperature 1Md polymorph was found to be younger than the enclosing rock, whereas high-temperature (2M) muscovite was distinctly older. This implies that the 1Md material is diagenetic (later diagenetic) and the 2M material detrital. Unfortunately, the 1Md material is finer grained and may leak Ar more rapidly, thus giving it an erroneously young age.

The best method of documenting clay-mineral transformations during later diagenesis is by study of sediments which have been subjected to known amounts of burial. This has been done by Burst (1959), who studied samples of the Eocene Wilcox formation from the Gulf Coastal Plain. This formation was chosen because (1) the top of the formation is a time line; (2) it dips basinward so that samples can be taken from the outcrop to depths greater than 16,000 feet; (3) the assumption of a same source for the entire formation during original deposition (the ancestral Mississippi River) is reasonable; (4) there has been little tectonism in the area. Burst found that from a depth of 0 to 1000 meters the principal clay is montmorillonite and that it has undergone little or no change with depth. From 1000 to 3500 meters the expandability of the montmorillonite decreases, and below 4000 meters only a small proportion of the clay shows expansion upon treatment with ion-solvating liquids. This progressive change in expansion characteristics with depth was interpreted to mean that montmorillonite is converted during later diagenesis to a mixed-layer illite–montmorillonite, and with deep enough burial, to illite without interlayer montmorillonite. Burst also cites an increase in the degree of crystallinity (or amount) of chlorites in the deeper samples as evidenced by a sharpening of x-ray diffraction peaks.

Other workers have also described a decrease of montmorillonite with burial to thousands of meters. Among them are Quaide (1957), Speights and Brunton (1961), Kubler (1964), and Perry (1969). The work of Perry (1969) is of special interest. Like Burst, Perry studied the changes in clay mineralogy as a function of burial depth (1000 to 5500 meters) in the Gulf Coast. However, along with x-ray mineralogical identifications, he performed chemical analyses and obtained *in situ* temperature data. In addition, micropaleontological identifications enabled the ascertainment of the position of

---

* Polymorphic forms of mica are based on different modes of vertical stacking of the 10-Å unit cell layers (Yoder and Eugster, 1955).

the original depositional site relative to the shoreline (i.e., brackish, neritic, etc.). Perry confirmed Burst's finding of decreasing expandability of montmorillonite with depth. He showed, conclusively, that this was due to the formation of interstratified illite in the montmorillonite structure. In all cases a monotonic decrease with depth in the percent of expandable montmorillonite layers was found which correlated well with increases in percent K_2O in the fine fraction of the sediment (<0.5 to 1.0 micron). This is shown in Fig. 9-8. The rate of change with depth of percent montmorillonitic layers was found to be controlled primarily by the temperature gradient and varied as the gradient varied. At about 30 percent montmorillonite the interlayering became ordered. Disagreements with Burst include the observation that chlorite and discrete-phase illite did not consistently increase with depth, and where they did, the increases could be attributed to a change in the original detrital clay mineralogy brought about by changing distances from the original shoreline.

Perry also found that the percentage of montmorillonite interlayers decreased to a limiting value of about 20 percent and did not drop further with increasing temperature (see Fig. 9-9). This suggests some sort of stability for the mixed-layer mineral. The main source of potassium for the illitic interlayers apparently was coarser-grained detrital illite, or mica in the sediment. This was indicated by a slight drop in percent of mica with depth, and by the fact that the whole rock potassium content remained constant during diagenesis. In other words, the sediment acted as a closed system with respect to potassium. Since mica is the only source of potassium in the rock abundant enough to be able to provide the K^+ necessary for forming interlayered illite under closed system conditions, it must have been the principal source of potassium. This, along with the leveling off of montmorillonite at 20 percent,

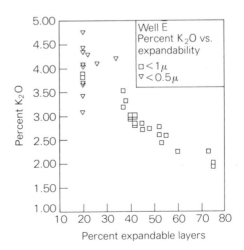

Fig. 9-8 Plot of weight percent K_2O in the clay fraction versus percent expandable (montmorillonite) layers for buried sediments from the Gulf Coast. Scatter for values at high percent K_2O is due to variable amounts of kaolinite, which acts as a K_2O-free dilutant, in the samples. For lower K_2O values, kaolinite is present in roughly constant concentrations. (*After Perry and Hower, 1970.*)

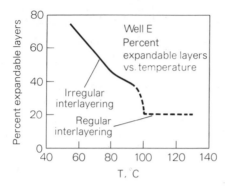

Fig. 9-9   Plot of percent montmorillonite (expandable) layers versus temperature for a drill hole from the Gulf Coast. Temperatures are *in situ* measured values which increase monotonically with depth. (*After Perry and Hower, 1970.*)

leads to a problem.   It means that fine-grained illite present as an interlayer in mixed-layer clay at concentrations <80 percent is more stable at diagenetic temperatures than coarser-grained mica.   This is in contradiction to the conclusions of Burst (1959) and most others, who have assumed that pure illite was an expected end product of the later diagenesis of montmorillonite. Certainly at higher temperatures characteristic of low-grade metamorphism the stable layer-type silicates are well-crystallized mica and chlorite.   Thus, if mixed-layer montmorillonite–illite has a thermodynamic stability field, it is confined to lower temperatures.   This possibility is intriguing and it is hoped that further work will help to elucidate the thermodynamic stability of mixed-layer clays of this type.

REFERENCES

Altschuler, Z. S., Dwornik, E. J., and Kramer, H., 1963, Transformation of montmorillonite to kaolinite during weathering, *Science*, v. 141, pp. 148–152.

Arrhenius, G., 1963, Pelagic sediments, in Hill, M. N. (ed.), *The sea: ideas and observations*, v. 3, Interscience-Wiley, New York, pp. 655–727.

Beall, A. O., and Fischer, A. G., 1969, Sedimentology, in *Initial reports of the Deep Sea Drilling Project (JOIDES)*, v. I, U.S. Government Printing Office, pp. 521–593.

Bear, F. E., 1964, *Chemistry of the soil*, 2d ed., Reinhold, New York, 515 p.

Biscaye, P. E., 1965, Mineralogy and sedimentation of recent deep-sea clay in the Atlantic Ocean and adjacent seas and oceans, *Geol. Soc. Am. Bull.*, v. 76, pp. 803–832.

Brown, G., 1961, *The X-ray identification and crystal structures of clay minerals*, Mineralogical Soc., London, 544 p.

Burst, J. F., 1959, Post-diagenetic clay mineral environmental relationships in the Gulf Coast Eocene, *Clays and clay minerals*, 6th Natl. Conf., Pergamon, New York, pp. 327–341.

Calvert, S. E., 1966, Origin of diatom-rich sediments from the Gulf of California, *Jour. Geol.*, v. 74, pp. 546–565.

Carroll, D., and Starkey, H. C., 1960, Effect of sea water on clay minerals, in Swineford, A. (ed.), *Clays and clay minerals*, 7th Natl. Conf., Pergamon, New York, pp. 80–101.

Dasch, E. J., 1969, Strontium isotopes in weathering profiles, deep-sea sediments, and sedimentary rocks, *Geochim. et Cosmochim. Acta*, v. 33, pp. 1521–1552.

————, Hills, F. A., and Turekian, K. K., 1966, Strontium isotopes in deep-sea sediments, *Science*, v. 153, pp. 295–297.

Deffeyes, K. S., 1965, *The Columbia River flood and the history of the oceans*, Pacific NW Oceanographers Annual Meeting, Corvallis, Oregon.

Emery, K. O., 1960, *The sea off southern California*, Wiley, New York, 365 p.

Eugster, H. P., and Jones, B. F., 1968, Gels composed of sodium-aluminum silicate, Lake Magadi, Kenya, *Science*, v. 161, pp. 160–163.

Foster, M. D. 1954, The relation between illite, beidellite, and montmorillonite, in Swineford, A., and Plummer, N. (eds.), *Clays and clay minerals*, 2d Natl. Conf. on Clays and Clay Minerals, Natl. Acad. Sci. Pub. 327, pp. 386–397.

Frondel, C., 1962, *The system of mineralogy, v. III, Silica minerals*, 7th ed., Wiley, New York, 334 p.

Garrels, R. M., 1967, Genesis of some ground waters from igneous rocks, in Abelson, P. H., *Researches in geochemistry*, v. 2, Wiley, New York, pp. 405–420.

————, and Christ, C. L., 1965, *Solutions, minerals, and equilibria*, Harper, New York, 450 p.

Goldich, S. S., 1938, A study in rock weathering, *Jour. Geol.*, v. 46, pp. 17–58.

Griffin, G. M., 1962, Clay mineral facies—products of weathering intensity and current distribution in the northeastern Gulf of Mexico, *Geol. Soc. Am. Bull.*, v. 73, pp. 737–768.

Grim, R. E., 1968, *Clay mineralogy*, 2d ed., McGraw-Hill, New York, 596 p.

Hower, J., Hurley, P. M., Pinson, W. H., and Fairbairn, H. W., 1963, The dependence of K-Ar on the mineralogy of various particle size ranges in a shale, *Geochim. et Cosmochim. Acta*, v. 27, pp. 405–410.

Hurley, P. M., Heezen, B. C., Pinson, W. H., and Fairbairn, H. W., 1963, K-Ar age values in pelagic sediments of the North Atlantic, *Geochim. et Cosmochim. Acta*, v. 27, pp. 393–399.

Iler, R. K., 1955, *The colloid chemistry of silica and silicates*, Cornell, Ithaca, New York, 324 p.

Jackson, M. L., 1964, Chemical composition of soils, in Bear, F. E., *Chemistry of the soil*, Reinhold, New York, pp. 71–141.

Johns, W. D., and Grim, R. E., 1958, Clay mineral composition of Recent sediments from the Mississippi River delta, *Jour. Sed. Petrology*, v. 28, pp. 186–199.

Keller, W. D., 1957, *The principles of chemical weathering*, Lucas Bros., Columbia, Mo., 111p.

————, 1963, Diagenesis of clay minerals: a review, *Clays and clay minerals*, Proc. 11th Natl. Conf., Pergamon, New York, pp. 136–157.

Kononova, M. M., 1966, *Soil organic matter*, 2d Eng. ed., Pergamon, New York, 544 p.

Krauskopf, K. B., 1956, Dissolution and precipitation of silica at low temperatures, *Geochim. et Cosmochim. Acta*, v. 10, pp. 1–26.

Kubler, B., 1964, Les argiles, indicateurs de metamorphisme, *Revue de l'Institut Français du Petrole et Annales des Combustibles Liquides*, v. 19, pp. 1093–1112.

Livingstone, D. A., 1963, Chemical composition of rivers and lakes, *U.S. Geol. Surv. Prof. Paper 440-G*, 64 p.

Mackenzie, F. T., and Garrels, R. M., 1966, Chemical balance between rivers and oceans, *Am. Jour. Sci.*, v. 264, pp. 507–525.

Marshall, C. E., 1964, *The physical chemistry and mineralogy of soils*, Wiley, New York, 388 p.

Mohr, E. C. J., and van Baren, F. A., 1954, *Tropical soils*, Interscience, New York, 498 p.

Morey, G. W., Fournier, R. O., and Rowe, J. J., 1962, The solubility of quartz in water in the temperature interval from 25° to 300°C, *Geochim. et Cosmochim. Acta*, v. 26, pp. 1029–1043.

Parham, W. E., 1966, Lateral variations of clay mineral assemblages in modern and ancient sediments, *Proc. of Intl. Clay Conf.*, Israel Program for Scientific Translations, Jerusalem.

Parks, G. A., 1965, The isoelectric points of solid oxides, solid hydroxides, and aqueous hydroxo complex systems, *Chem. Rev.*, v. 65, pp. 177–197.

Perry, E. A., 1969, Burial diagenesis in Gulf Coast pelitic sediments, Ph.D. Thesis, Case Western Reserve University, Cleveland, Ohio, 121 p.

———, and Hower, J., 1970, Burial diagenesis in Gulf Coast pelitic sediments, *Clays and Clay Minerals*, v. 18, no. 3.

Peterson, M. N. A., and Griffin, J. J., 1964, Volcanism and clay minerals in the southeastern Pacific, *Jour. Marine Research*, v. 22, pp. 13–21.

———, and von der Borch, C. C., 1965, Chert: modern inorganic deposition in a carbonate-precipitating locality, *Science*, v. 149, pp. 1501–1503.

Quaide, W., 1957, Clay minerals from the Ventura Basin, *Jour. Sed. Petrology*, v. 27, pp. 336–341.

Rich, C. I., and Kunze, G. W., 1964, *Soil clay mineralogy*, University of North Carolina Press, Chapel Hill, North Carolina, 330 p.

Ross, C. S., and Hendricks, S. B., 1945, Minerals of the montmorillonite group, *U.S. Geol. Surv. Prof. Paper 205-B*, pp. 23–47.

Russell, K. L., 1970, Geochemistry and halmyrolysis of clay minerals, Rio Ameca, Mexico, *Geochim. et Cosmochim. Acta*, v. 34, pp. 893–907.

Sherman, G. D., 1952, The genesis and morphology of the alumina-rich laterite clays, Am. Inst. Min. Metal. Eng., *Problems in clay and laterite genesis*, pp. 154–161.

Siever, R., 1962, Silica solubility, 0–200°C, and the diagenesis of siliceous sediments, *Jour. Geol.*, v. 70, pp. 127–150.

———, 1971, Low temperature geochemistry of silicon, in Wedepohl, K. H. (ed.), *Handbook of geochemistry*, v. II-1, Springer-Verlag, Berlin, section 14.

———, Beck, K. C., and Berner, R. A., 1965, Composition of interstitial waters of modern sediments, *Jour. Geol.*, v. 73, pp. 39–73.

Speights, D. B., and Brunton, G., 1961, Clay mineral distribution in Permo-Pennsylvanian shales of Val Verde Basin and Yates-Todd Arch, Texas, *Am. Assoc. Petroleum, Geologists Bull.*, v. 45, pp. 1957–1970.

Sverdrup, H. U., Johnson, M. W., and Fleming, R. H., 1942, *The oceans, their physics, chemistry and biology*, Prentice-Hall, Englewood Cliffs, N.J., 1087 p.

Turekian, K. K., 1968, The oceans, streams, and atmosphere, in Wedepohl, K. H. (ed.), *Handbook of geochemistry*, v. I, Springer-Verlag, Berlin, pp. 297–323.

Velde, B., 1965, Experimental determination of muscovite polymorph stabilities, *Am. Mineralogist*, v. 50, pp. 436–449.

———, and Hower, J., 1963, Petrological significance of illite polymorphism in Paleozoic sedimentary rocks, *Am. Mineralogist*, v. 48, pp. 1239–1254.

Walton, H. F., 1959, Ion exchange equilibria, in Nachod, F. C. (ed.), *Ion exchange, theory and practice*, Academic, New York, pp. 3–28.

Weaver, C. E., 1958, The effects and geologic significance of potassium "fixation" by expandable clay minerals derived from muscovite, biotite, chlorite, and volcanic material, *Am. Mineralogist*, v. 43, pp. 839–861.

———, 1967, Potassium, illite, and the ocean, *Geochim. et Cosmochim. Acta*, v. 31, pp. 2181–2196.

Whitehouse, U. G., and McCarter, R. S., 1958, Diagenetic modification of clay mineral types in artificial sea water, *Clays and Clay Minerals*, Natl. Acad. Sci. Publ. 566, pp. 81–119.

Wollast, R., 1967, Kinetics of the alteration of K-feldspar in buffered solutions at low temperature, *Geochim. et Cosmochim. Acta*, v. 31, pp. 635–648.

————, Mackenzie, F. T., and Bricker, O. P., 1968, Experimental precipitation and genesis of sepiolite at Earth surface conditions, *Am. Mineralogist*, v. 53, pp. 1645–1662.

Yoder, H. S., and Eugster, H. P., 1955, Synthetic and natural muscovites, *Geochim. et Cosmochim. Acta*, v. 8, pp. 225–280.

10
Diagenesis of Iron Minerals

The common iron minerals which form under sedimentary conditions (see Table 10-1) include hematite, goethite, siderite, glauconite, and pyrite (plus its precursors, mackinawite and greigite). Dolomite may also contain appreciable iron but not enough to be considered as an iron mineral. Hematite and goethite are the oxidation products of the weathering of ferrous minerals and constitute a major source of detrital iron in sediments. By contrast glauconite, siderite, and the iron sulfides form only during diagenesis.

MINERAL STABILITY

The purpose of this section is to discuss the relative stability under sedimentary conditions of the principal iron minerals. Since iron exhibits two oxidation states, Fe^{++}, and Fe^{3+}, stability must be a function of the redox state of the system, which can be expressed in terms of Eh. In addition, inclusion of several different elements in the composition of these minerals necessitates the use of numerous additional parameters such as pH, dissolved CO_2, etc. The usual method of representation is by means of Eh-pH diagrams for constant total dissolved carbon, sulfur, and silica (Garrels and Christ, 1965). The diagrams

Table 10-1   Common sedimentary iron minerals of low-temperature origin discussed in this chapter

Mineral	Formula	Color imparted to sediments or rocks
Hematite	Fe_2O_3	Red to purple
Goethite	$HFeO_2$	Yellow to brown
Siderite	$(Fe,Mg)\,CO_3$	None
Glauconite	$K\text{-}Fe^{+3}\text{-}Fe^{+2}\text{-}$ hydrous aluminosilicate	Green or none
Pyrite; marcasite	FeS_2	Light gray or none*
Greigite	Fe_3S_4	Black to gray*
Mackinawite	$Fe_{1+x}S$	Black to gray*

* Colors in recent sediments only.   In ancient rocks the black and gray colors are due to carbonized organic matter.

are constructed by plotting equations, calculated from thermodynamic data, for equilibrium between two minerals as boundary curves.   The dissolved carbon and sulfur species, since they vary in abundance as a function of Eh and pH (see Chapter 7, Fig. 7-1), require the use of several equations to describe equilibrium between each mineral pair.   One disadvantage of the application of Eh-pH diagrams to sediments is that the pH of most sediments is roughly constant.   This is especially true of marine sediments where the pH is buffered by $CaCO_3$ or by $HCO_{3\,aq}^-$ to values ranging only from about 7.0 to 8.0 (Baas Becking et al., 1960).   In addition, redox equilibrium between all dissolved species, which is a necessary assumption in constructing Eh-pH diagrams for iron minerals, is not approached closely in natural waters.   This has been shown in Chapter 7.   Since the concentration and distribution of dissolved carbon and sulfur species is a function of bacterial activity which is not always predictable, it is best to consider these species as independent variables and not dependent variables as assumed in an Eh-pH diagram.

Because of the above problems an alternate form of relative stability representation has been adopted in this chapter.   The variables chosen are Eh, P_{CO_2}, and pS^{--}.   The parameter pS^{--} is the negative logarithm of the activity of sulfide ion and can be measured directly with an electrode.   For marine sediments equilibrium with $CaCO_3$ at a constant value of $a_{Ca^{++}}$ is assumed.   Thus, pH is directly proportional to $\log P_{CO_2}$.   This is shown as follows:

$$CO_{2\,gas} + H_2O + Ca_{aq}^{++} \leftrightharpoons CaCO_3 + 2H_{aq}^+$$

$$K = 10^{-9.75} = \frac{a_{H^+}^2}{P_{CO_2}\,a_{Ca^{++}}} \tag{1}$$

For sea water $a_{Ca^{++}} = 10^{-2.58}$ (Chapter 3).   Therefore, for sea water

$$pH = 6.17 - \tfrac{1}{2} \log P_{CO_2} \tag{2}$$

Because of the limited range of pH in marine sediments, P_{CO_2} is either held constant or varied over just three orders of magnitude ($10^{-1.0}$ to $10^{-4.0}$) on the diagrams.   The assumptions of constant $a_{Ca^{++}}$ and variation of P_{CO_2} within this range are reasonable approximations for most marine sediments.

Glauconite is not included on the diagrams because of a lack of thermo-dynamic data.   Also not included are goethite, mackinawite, and greigite because of their instability relative to hematite, pyrrhotite, and pyrite. Magnetite and pyrrhotite are included because authigenic pyrrhotite is occasionally found in sediments, and magnetite is ubiquitous as a detrital mineral.   Equilibria and corresponding equations, using data for ΔF_f° from Appendix I, for 25°C are

Siderite-hematite

$$2FeCO_{3 \, siderite} + H_2O \leftrightharpoons Fe_2O_{3 \, hematite} + 2CO_{2 \, gas} + 2H_{aq}^+ + 2e$$

$$Eh = 0.286 - 0.0592 \, pH + 0.0592 \log P_{CO_2} \tag{3}$$

From (2)

$$Eh = -0.079 + 0.0888 \log P_{CO_2} \tag{4}$$

Magnetite-hematite

$$2Fe_3O_{4 \, magnetite} + H_2O \leftrightharpoons 3Fe_2O_{3 \, hematite} + 2H_{aq}^+ + 2e$$

$$Eh = 0.221 - 0.0592 \, pH \tag{5}$$

From (2)

$$Eh = -0.144 + 0.0296 \log P_{CO_2} \tag{6}$$

Siderite-magnetite

$$3FeCO_{3 \, siderite} + H_2O \leftrightharpoons Fe_3O_{4 \, magnetite} + 3CO_{2 \, gas} + 2H_{aq}^+ + 2e$$

$$Eh = 0.319 - 0.0592 \, pH + 0.0888 \log P_{CO_2} \tag{7}$$

From (2)

$$Eh = -0.046 + 0.1184 \log P_{CO_2} \tag{8}$$

Siderite-pyrite

$$FeCO_{3 \, siderite} + 2H_{aq}^+ + 2S_{aq}^{--} \leftrightharpoons FeS_{2 \, pyrite} + CO_{2 \, gas} + H_2O + 2e$$

$$Eh = -1.56 + 0.0592 \, pH + 0.0592 \, pS^{--} + 0.0296 \log P_{CO_2} \tag{9}$$

From (2)

$$Eh = -1.19 + 0.0592 \, pS^{--} \tag{10}$$

Magnetite-pyrite

$$Fe_3O_{4 \, magnetite} + 6S_{aq}^{--} + 8H_{aq}^{+} \leftrightharpoons 3FeS_{2 \, pyrite} + 4H_2O + 4e$$

$$Eh = -2.50 + 0.1184 \, pH + 0.0888 \, pS^{--} \tag{11}$$

From (2)

$$Eh = -1.77 + 0.0888 \, pS^{--} - 0.0592 \log P_{CO_2} \tag{12}$$

Pyrrhotite-pyrite

$$FeS_{pyrrhotite} + S^{--} \leftrightharpoons FeS_{2 \, pyrite} + 2e$$

$$Eh = -0.782 + 0.0296 \, pS^{--} \tag{13}$$

Hematite-pyrite

$$Fe_2O_{3 \, hematite} + 4S_{aq}^{--} + 6H_{aq}^{+} \leftrightharpoons 2FeS_{2 \, pyrite} + 3H_2O + 2e$$

$$Eh = -3.41 + 0.1776 \, pH + 0.1184 \, pS^{--} \tag{14}$$

From (2)

$$Eh = -2.31 - 0.0888 \log P_{CO_2} + 0.1184 \, pS^{--} \tag{15}$$

Pyrrhotite-magnetite

$$3FeS_{pyrrhotite} + 4H_2O \leftrightharpoons Fe_3O_{4 \, magnetite} + 8H_{aq}^{+} + 3S_{aq}^{--} + 2e$$

$$Eh = 2.66 - 0.2368 \, pH - 0.0888 \, pS^{--} \tag{16}$$

From (2)

$$Eh = 1.20 + 0.1184 \log P_{CO_2} - 0.0888 \, pS^{--} \tag{17}$$

Pyrrhotite-siderite

$$FeS_{pyrrhotite} + CO_{2 \, gas} + H_2O \rightleftharpoons FeCO_{3 \, siderite} + 2H_{aq}^{+} + S_{aq}^{--}$$

$$2pH + pS^{--} + \log P_{CO_2} = 26.4 \tag{18}$$

From (2)

$$pS^{--} = 14.1 \tag{19}$$

Two diagrams (Figs. 10-1 and 10-2) are plotted, one for nonsulfidic sediments where pS^{--} is so high that pyrite and pyrrhotite do not plot on a Eh-P_{CO_2} diagram, and the other for sulfidic sediments where $\log P_{CO_2}$ is assumed to be -2.40 (corresponding to pH = 7.37) throughout the Eh-pS^{--} diagram.

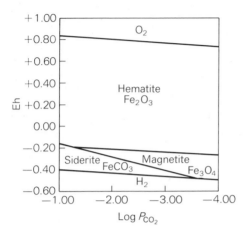

Fig. 10-1 Eh-log P_{CO_2} diagram for hematite, magnetite, and siderite in marine sediments. $T = 25°C$, $P_{total} = 1$ atm, $a_{Ca^{++}} = 10^{-2.58}$ equilibrium with calcite assumed. O_2 and H_2 represent areas where water is thermodynamically unstable relative to the respective gases. The value for pS^{--} is assumed to be so high that pyrite and pyrrhotite do not plot stably.

Because of the problems discussed in Chapter 7, measured electrode values of Eh in sediments are not accurate guides to thermodynamic Eh as plotted in Figs. 10-1 and 10-2. Nevertheless, experimental and natural measurements are in overall qualitative agreement with the mineralogical predictions of the diagrams. In aerobic, high Eh sediments, only ferric oxides are found. Siderite forms only at low Eh, high pS^{--} and high P_{CO_2},

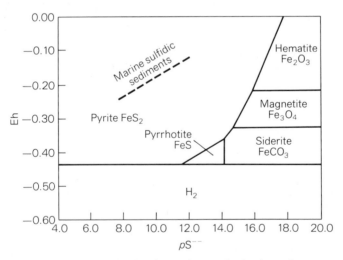

Fig. 10-2 Eh-pS^{--} diagram for pyrite, pyrrhotite, hematite, magnetite, and siderite. pH = 7.37, $\log P_{CO_2} = -2.40$. $T = 25°C$, $P_{total} = 1$ atm. The field marked H_2 corresponds to the area where water is unstable relative to H_2 gas. Measurements of natural sulfidic marine sediments fall closely along the dashed line. (*After Berner, 1964b.*)

and pyrite at low Eh and moderate to low pS^{--}. Measured values of Eh and pS^{--} for marine pyritic sediments are plotted in Fig. 10-2 and as predicted, fall within the stability field of pyrite. The diagrams are therefore useful and may be used (with some qualifications) to decipher the chemical conditions present during the formation of authigenic iron minerals in marine sedimentary rocks. The major error is the omission of glauconite. Since glauconite is a $Fe^{3+} + Fe^{++}$ mineral it would probably plot in the same general region as magnetite, but positions of the actual boundaries and the size of the stability field remain unknown.

RED BED FORMATION

Fine-grained goethite, $HFeO_2$, is a very common constituent of the weathering product limonite, which is the chief yellow-brown pigment in sediments. Although abundant in modern sediments and on weathered outcrops, limonite is rare in buried ancient sedimentary rocks (Fischer, 1963); therefore, assuming uniformitarianism, limonite must disappear during diagenesis. If all ancient rocks were sideritic, glauconitic, or pyritic, the disappearance could be readily explained. However, hematite is a common constituent of sedimentary rocks and often imparts a red coloration to rocks, giving rise to the term *red beds*. If limonite disappears during diagenesis, some of it may be dehydrated to hematite. In order for this to happen, rather than reduction, the original sediment would have to be relatively free of decomposable organic matter so as to maintain a high enough Eh to stabilize hematite (Fig. 10-1). Otherwise, the iron would be reduced by the action of iron-reducing bacteria, organic reducing agents, or H_2S (Berner, 1970b). The fact that ancient red beds are very low in organic carbon is in agreement with this prediction. Also, because organic matter is generally rather abundant in epicontinental marine sediments, almost all red beds are nonmarine.

Thermodynamically, the transformation of limonitic goethite to hematite during diagenesis is feasible. From studies of differential solubility in HCl solution, the author (Berner, 1969) has found that for the reaction

$$2HFeO_{2\,goethite} \rightarrow Fe_2O_{3\,hematite} + H_2O_{liq}$$

the value of $\Delta F°$ at 25°C is −0.66 kcal/mole and at 0°C is −0.40 kcal/mole if goethite is present in crystal sizes on the order of hundreds of angstroms. This conclusion is in qualitative agreement with the work of Langmuir (1970), who has calculated the effect of surface energy on goethite dehydration. From Eq. (86) of Chapter 2, $\Delta F°$ at 25°C and $a_{H_2O} = 1$ for the above reaction is given by

$$\Delta F° = \Delta F_\infty° - \frac{4\sigma_g v_g B_g'}{3r_g} + \frac{2\sigma_h v_h B_h'}{3r_h} \qquad (20)$$

where the subscripts g and h refer to goethite and hematite, respectively, and ΔF°_∞ refers to standard states of coarse crystallinity. From Langmuir's estimates $\Delta F^\circ_\infty = 0.54$ kcal/mole, $\sigma_g = 1600$ ergs/cm^2, $\sigma_h = 1200$ ergs/cm^2. If $B'_g = 10$ (laths) and $B'_h = 6$ (cubes), the conversion of typical limonitic goethite, with a lath width r_g of 500 Å, to the moderately well-crystallized hematite characteristic of ancient red beds ($r_h > 5000$ Å) takes place with $\Delta F^\circ = -1.8$ kcal/mole. With an increase in temperature or a decrease in a_{H_2O}, hematite is further stabilized. Since temperature increases during burial, it is apparent that limonitic goethite has essentially no geochemical stability field and persists in sediments because of slow rates of interconversion between highly insoluble oxides at low temperature (Garrels, 1959).

In contrast to the above conclusions, earlier experimental phase-equilibria work (e.g., Smith and Kidd, 1949; Schmalz, 1959) had suggested that the stable iron oxide in aqueous solution at most diagenetic temperatures was goethite, and that hematite became stable only at high temperature (>130°C) or at very low activities of water. Unfortunately, in none of the earlier studies was reversible equilibrium demonstrated; goethite was dehydrated to hematite, but hematite was never hydrated to goethite.* Nevertheless, these results have been used to bolster two different paleoclimatic theories for the origin of red beds. Both theories are based on the reasoning that, if goethite is stable relative to hematite, then most red beds must have been deposited originally as red, hematitic sediments. The red color is derived either by the erosion and deposition of red soils (e.g., Krynine, 1950) or by the dehydration of limonite to hematite in dry air or in extremely saline brines (e.g., Schmalz, 1959). Since red soils are forming today only at lower latitudes in warm, humid climates, proponents of the red soil theory believe that many red beds originated under subtropical climates. By contrast, the common association of red beds with evaporites lends credence to the idea that red beds are formed at low relative humidity and, thus, are an indicator of arid climates.

The data of Langmuir and the author show that hematite requires no special conditions in order to form from limonite during diagenesis so that many red beds may have been originally yellow and not red. This agrees with the observations of Walker (1967) that transformation of limonite to hematite in Recent and Pleistocene sediments actually takes place at low temperatures in dilute ground waters. Thus, it might appear that red beds are not good paleoclimatic indicators. However, the red color does indicate that ferric oxides, whether originally yellow or red, have withstood reduction to ferrous minerals during diagenesis because of a lack of metabolizable organic matter in the sediments. This means that before burial essentially all organic matter

---

* Smith and Kidd incorrectly described the aging of freshly precipitated red ferric hydrate gel to goethite as the conversion of hematite to goethite.

was destroyed by bacterial processes, a common attribute of tropical climates; or that there was little vegetation at the site of deposition, an attribute of arid climates. Red beds may have some climatic significance, but it is not because they are red; it is because they contain ferric, rather than ferrous, authigenic iron minerals, a factor which has been emphasized by Van Houten (1968).

SIDERITE FORMATION

The conditions under which siderite is thermodynamically stable are severely restricted. Figures 10-1 and 10-2 show that Eh must be low and pS^{--} must be high, which is an unexpected combination for marine sediments. This is because low Eh is the result of the anaerobic bacterial decomposition of organic matter, and in sea water, which contains abundant dissolved sulfate, anaerobic decomposition almost always includes the reduction of sulfate to H_2S. If sulfate reduction enables the attainment of thermodynamically reversible redox equilibrium between $SO_{4\,aq}^{--}$ and HS_{aq}^{-}, or H_2S_{aq}, then siderite has no stability field in marine sediments. This can be seen by comparing Fig. 10-2 with Fig. 7-1 of Chapter 7. For Eh values where siderite is stable, almost all the total dissolved sulfur at equilibrium must be present as either H_2S_{aq} or HS_{aq}^{-}; i.e., sulfate reduction must be essentially complete. Thus, siderite can form from sea water only where sulfate reduction is inhibited so that metastable $SO_{4,aq}^{--}$ persists. This situation is unusual and as a result, marine siderite is rare. It has never been observed forming in modern marine sediments.

By comparison, siderite is a relatively common constituent of ancient nonmarine sediments where it is normally found in association with coal beds and fresh-water clays. This association is a predictable result of thermodynamic stability. Some fresh waters are low in dissolved sulfate so that the anaerobic bacterial decay of organic matter may enable the attainment of a low Eh and high P_{CO_2} without the formation of appreciable H_2S. Since siderite is a carbonate, it is soluble under acidic conditions and does not form in mineral-free swamps and bogs. However, many lake sediments and some bogs are not acidic because of buffering by fine-grained silicate minerals and/or a high carbonate alkalinity (Baas Becking et al., 1960). Thus, siderite could form in a clayey bog or lake sediment, low in dissolved sulfate and high in organic matter. The result after lithification would be a coaly sideritic-clay ironstone, the type of primary occurrence commonly found in sedimentary rocks.

Evidence for the formation of siderite in modern nonmarine sediments is needed to check out the predictions based on thermodynamic stability. It may be much more common than is realized. This is partly because there are no characteristic features, such as a pigmenting color, which enable recognition in the field, but also because it is a minor constituent which, if fine-grained and dispersed in a sediment (or rock), can be easily missed during x-ray, chemical, or optical study. *Indirect* evidence for the formation of siderite in the anaerobic muds of a fresh-water pond (Doyle, 1967) is provided by analyses of the pore waters, which indicate supersaturation with respect to $FeCO_3$.

Siderite is normally recognized in ancient rocks because it forms prominent concretions. The concretions are postdepositional, and they may occur in marine beds, which has led to the assumption that precipitation took place in interstitial sea water. However, from what has been said above, it is quite possible that the concretions have formed as a result of the exposure of marine sediments, after uplift, to nonmarine anaerobic ground waters.

In addition to waters containing H_2S, siderite also cannot form in a water rich in dissolved calcium. For the reaction

$$FeCO_{3 \text{ siderite}} + Ca_{aq}^{++} \leftrightarrows CaCO_{3 \text{ calcite}} + Fe_{aq}^{++}$$

$$K = 0.05 = \frac{a_{Fe^{++}}}{a_{Ca^{++}}} \tag{21}$$

For siderite to be stable relative to calcite the concentration of iron must be greater than 5 percent that of calcium. In sea water and marine sediments it is less than 0.1 percent (Berner, 1970b). Therefore, for this reason alone, siderite is thermodynamically unstable in sea water. In low-calcium anaerobic fresh water, dissolved ferrous iron is often much higher relative to calcium so that here siderite can be more stable than calcite. A good example is the sediment pore waters of Linsley Pond, Connecticut (Doyle, 1967), where Fe_{aq}^{++} is present in *higher* concentrations than Ca_{aq}^{++}.

GLAUCONITE FORMATION

Free-energy data for glauconite are not available; thus its stability is not known. Glauconite is also a poorly defined phase. The term is applied to green globular aggregates in sediments which may consist of (Burst, 1958):

I. Well-crystallized high-potassium, mica structure (true mineral glauconite)
II. Disordered, low-potassium, mica structure
III. Extremely disordered, low-potassium, expandable, montmorillonite structure
IV. Mixtures of two or more clay minerals unrelated to true mineral glauconite

True mineral glauconite is a potassium iron aluminosilicate with a high Fe^{3+}/Fe^{++} ratio. It is normally found only in Paleozoic rocks. Modern glauconites tend to be types III and IV (Bell and Goodell, 1967; Hower, 1961). Hower believes that during long-term diagenesis, type I, or mineral, glauconite is formed from types II, III, and IV. This process is reminiscent of the formation of illite during deep diagenesis as discussed in Chapter 9.

The occurrences in modern sediments (Cloud, 1955) provide some idea regarding the environment of formation of glauconite. It is almost strictly marine and is most abundant in areas with low rates of deposition such as current-swept bank tops. This suggests that it forms slowly at the sediment-water interface. It is generally associated with organic matter (e.g., as foram shell fillings) but in an overall aerobic environment. It does not form in sulfidic sediments where pyrite is presumably more stable. The occurrence with both organic matter and dissolved oxygen, formation at the sediment-water interface, and high Fe^{3+}/Fe^{++} ratio all suggest that glauconite is stable in an environment of intermediate, and probably fluctuating, redox potential. Free-energy data, at least for mineral glauconite, are needed to test this hypothesis. Low-temperature glauconite synthesis, which is yet to be accomplished, would also help in elucidating its stability and mechanism of formation.

Before going on to a detailed discussion of sedimentary pyrite, a representation of the typical environment of formation for each diagenetic iron mineral covered in this chapter is presented in the form of an idealized cross section in Fig. 10-3. This diagram is intended to illustrate different environments and should not be interpreted too literally as representing the only places where each iron mineral may form.

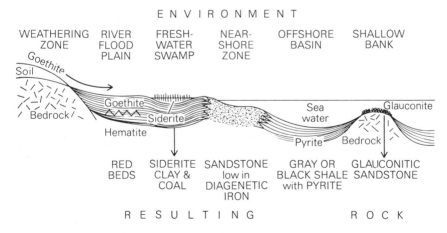

Fig. 10-3 Idealized cross section illustrating sedimentary environments where various diagenetic iron minerals are most likely to form. Detrital limonite and hematite are carried mainly as colloids or adsorbed coatings on fine-grained minerals (Carroll, 1958) and thereby are separated, by water turbulence in the near-shore zone, from sand grains poorer in reactive iron. This is why shales contain, on the average, more iron than do sandstones.

PYRITE* FORMATION

According to Fig. 10-2 pyrite is the stable iron mineral under low Eh and low pS^{--} conditions. Such an environment is characteristic of fine-grained marine sediments enriched in organic matter, where bacterial sulfate reduction is active (see Fig. 10-3). Organic sediments invariably contain some pyrite (or marcasite) and except for a few rare instances, it is always an authigenic mineral because detrital pyrite is chemically and physically very unstable. Discussion of the problem of sedimentary pyrite formation is here divided into three sections: sources of iron and sulfur, factors limiting formation, and mechanism of formation.

SOURCES OF IRON AND SULFUR

By far the chief source of iron for pyrite formation in most sediments is detrital iron minerals. Natural surface waters contain almost no dissolved iron, and biogenic materials are characteristically low in iron. In some sediments which are essentially free of detrital matter (e.g., carbonates) the principal source of iron can be that contained within organic remains, but as a result the pyrite content is low.

After deposition in an anaerobic environment detrital iron minerals can become solubilized by bacterial or inorganic processes. The most common process is reduction of ferric oxides to dissolved ferrous iron. Because of inhomogeneous solubilization, concentration gradients may occur and the dissolved ferrous iron may migrate to form concretions and other aggregations of pyrite (see Chapter 6). In this sense the direct source of iron is dissolved ferrous iron. However, the primary source is still detrital minerals added to the sediment.

Not all detrital iron minerals fully react to form pyrite. The most reactive phases are those that are very fine grained. This includes limonitic goethite and hematite and clay-sized chlorite. The fine-grained ferric oxides most commonly occur as adsorbed coatings on detrital grains, especially on clay minerals (Carroll, 1958). Sand- and silt-sized grains of magnetite and ferruginous silicates, by contrast, are relatively unreactive and are often found as metastable phases in otherwise pyritic sediments.

Pyrite forms as a result of the reaction of dissolved H_2S with iron minerals. The two major sources of H_2S are bacterial sulfate reduction and the decomposition of organic sulfur compounds derived from dead organisms. In most sediments dissolved sulfate is a far more important source. Proof of this contention is provided by many lines of evidence. For instance experimental studies (e.g., Skopintsev, 1961) have shown that during the bacterial decomposition of biogenic organic matter in sea water, increases in dissolved H_2S

---

* The term pyrite is used here in a generic sense to include both true pyrite and its dimorph, marcasite.

are equally matched by decreases in dissolved sulfate. There is little excess H_2S derived from organic sulfur compounds. Another line of evidence is that concentrations of pyrite sulfur in sediments often far exceed that which could be supplied by organic sulfur compounds. Marine sediments often contain more than 1 percent pyrite sulfur by dry weight. Since the organic sulfur content of marine organisms averages about 1 percent by dry weight (Kaplan et al., 1963), more than 100 percent organic matter would be required in many sediments if all the pyrite were derived from organic sulfur compounds. Actual concentrations of organic matter are much less, so dissolved sulfate must be the principal source of most of the pyrite sulfur.

In some lakes and swamps dissolved sulfate is very low and under such circumstances pyrite sulfur may emanate mainly from organic sulfur compounds. However, it should be noted that fresh-water sediments that are high in H_2S and/or pyrite generally occur in waters high in dissolved sulfate (Hutchinson, 1957).

FACTORS LIMITING PYRITE FORMATION

The three principal factors that limit the amount of pyrite or any other iron sulfide which may form in a sediment are the concentration and reactivity of iron compounds, the availability of dissolved sulfate, and the concentration of organic compounds which can be utilized by sulfate-reducing bacteria to produce H_2S. To simplify discussion the latter is henceforth referred to as metabolizable organic matter.

Probably the most important limiting factor is the concentration of metabolizable organic matter. All marine sediments originally contain some reactive iron compounds and dissolved sulfate. However, many contain no metabolizable organic matter and as a consequence no H_2S or pyrite forms. In organic-rich sediments the amount of pyrite may be directly controlled by organic material. This is a consequence of the reasonable assumption (see Chapter 7) that the rate of sulfate reduction in a sediment is directly proportional to the concentration of metabolizable organic matter. Higher rates of reduction resulting from a higher concentration of organic matter enable the attainment of higher concentrations of H_2S in pore waters in the presence of H_2S-removing processes such as air oxidation and diffusion to the overlying water. Because iron minerals exhibit a spectrum of reactivities toward H_2S, as outlined earlier, the higher the H_2S concentration in pore waters and the longer it is maintained, the greater should be the proportion of total detrital iron that is transformed to pyrite. Thus, if metabolizable organic matter were a limiting factor, one would expect to find a direct relation between the concentration of available organic matter and the fraction of total iron converted to pyrite. Such a relation is exhibited by the topmost portions of sediments from the central coast of Connecticut (see Fig. 10-4). It is probable that many other sediments also exhibit a similar relationship.

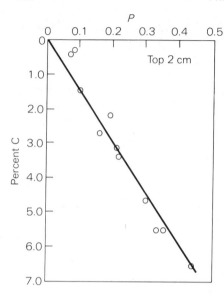

Fig. 10-4 A plot of percent total organic carbon versus the degree of pyritization of iron P for the top 2 cm of marine sediments from the central coast of Connecticut. (*After Berner, 1970a.*) $P = Fe_{pyrite}/Fe_{pyrite} + Fe_{HCl}$, where Fe_{HCl} refers to iron soluble in hot concentrated HCl solution. It is assumed that HCl-soluble iron includes all potentially reactive (toward H_2S) iron in the sediments and that metabolizable organic matter is a direct function of total organic carbon. For the uppermost portions of sediments in an areally restricted zone of sedimentation the latter assumption is probably reasonably valid.

In some sediments pyrite formation is not dependent upon the local content of metabolizable organic matter. In meromictic lakes or euxinic marine basins, where waters overlying the sediments contain dissolved H_2S, pyrite may form at, or even above, the sediment-water interface. Under these circumstances, because the H_2S concentration in bottom waters is relatively constant and areally uniform, the amount of pyrite added to the sediments may be controlled by the amount of reactive detrital iron delivered to the site of deposition. This, in turn, is a function of the circulation processes in the water mass which govern the distribution of fine-grained minerals containing reactive limonite, hematite, etc. In the Black Sea, which is the type locality and largest modern example of a euxinic basin, present-day pyrite formation is in agreement with the above reasoning and appears to be controlled mainly by the distribution of reactive detrital iron minerals (Ostroumov et al., 1961).

The concentration of total iron is seldom a limiting factor in pyrite formation. In almost all pyritic sediments a large excess of HCl-soluble iron is found, and since pyrite is insoluble in HCl, the iron must come from metastable detrital iron compounds which have not been transformed to pyrite. The reactivity of iron is much more important than its total concentration; at a given fixed concentration of H_2S the reactivity of the iron dictates what proportion will be transformed to pyrite. In some nonterrigenous sediments such as carbonates, the iron may be supplied almost entirely as highly reactive organic iron compounds. In this case the iron may be totally converted to pyrite, and thus may become a limiting factor.

In most marine sediments dissolved sulfate is not completely removed by bacteria in the upper zones where pyrite is formed.   As a result dissolved sulfate is usually not an important limiting factor in sedimentary pyrite formation.   Measurable concentrations of sulfate are commonly found several tens of centimeters below the depth where the degree of pyritization levels off to a constant value.   Occasionally highly organic sediments such as those affected by pollution are encountered where sulfate rapidly disappears with depth, but these sediments are atypical.   The reason why sulfate is not limiting is that it is able to continually diffuse into sediments from the overlying sea water.   As a result, concentrations of pyrite sulfur often exceed the value ($\sim$0.3 percent) expected for utilization of all originally buried sulfate. The sediment apparently acts as an open system with regard to sulfur.

MECHANISM OF FORMATION

Pyrite is not formed directly by the reaction of bacteriogenic sulfide with detrital iron minerals.   Instead other metastable iron sulfides crystallize first. This has been shown by laboratory experiments conducted so as to simulate sedimentary conditions (Berner, 1964a; Rickard, 1969).   The principal phases which initially form at room temperature and neutral pH by the reaction of H_2S and HS^- with fine-grained goethite or dissolved ferrous iron are mackinawite (tetragonal $Fe_{1+x}S$) and greigite (cubic Fe_3S_4).   Both are black and soluble in hot, concentrated HCl, whereas pyrite is not.

In recent sulfidic sediments black HCl-soluble iron sulfides are common. The exact nature of these phases is not well-known because of extreme difficulty in separating them from other sedimentary constituents.   However, in a few cases successful isolation has revealed that the black material consists of greigite, mackinawite, or both (Polushkina and Sidorenko, 1963; Berner, 1964a; Jedwab, 1967).   On the basis of the laboratory experiments it is probable that much of the unidentified black iron sulfide also consists of fine-grained mackinawite or greigite.   Very fine crystallite size is a consequence of rapid nucleation due to a high degree of supersaturation during precipitation (iron sulfides are very insoluble).   Because of their fine size the particles are adsorbed on larger grains as coatings which impart an overall black color to the sediment.

Greigite and mackinawite are thermodynamically metastable relative to pyrite and stoichiometric pyrrhotite.   From free-energy data of Appendix I, the following reactions can be shown to proceed with a free-energy change less than zero:

$$FeS_{mackinawite} \rightarrow FeS_{pyrrhotite}$$

$$Fe_3S_{4\ greigite} \rightarrow 2FeS_{pyrrhotite} + FeS_{2\ pyrite}$$

Thus, it would be expected that during diagenesis mackinawite and greigite should disappear.

In most recent marine sediments, the black iron sulfides do disappear and are transformed to pyrite. This is in agreement with equilibrium predictions based on measured values of Eh and pS^{--} (see Fig. 10-2). Evidence for transformation is shown by a decrease of acid-soluble FeS (e.g., see Fig. 10-5) and more simply by the usual disappearance of black color with depth (Van Straaten, 1954).

The mechanism of transformation is not well established but it is believed that the reactions

$$FeS + S^0 \rightarrow FeS_2$$

$$Fe_3S_4 + 2S^0 \rightarrow 3FeS_2$$

which proceed with a large change in free energy, are of considerable importance. The reaction of "FeS" with elemental sulfur has been invoked by Ostroumov

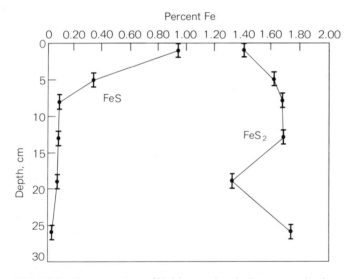

Fig. 10-5 Concentrations of FeS iron and pyrite iron versus depth for a tidal flat sediment from the central Connecticut coast. Loss of FeS with depth represents transformation to pyrite. A one-to-one correspondence between FeS decrease and FeS_2 increase, expected for closed-system transformation, is not present in tidal flat sediments. This is because FeS is also oxidized to limonite by dissolved O_2 occasionally stirred into the sediment by burrowing organisms and tidal currents. After bacterial removal of O_2, FeS is re-formed by bacteriogenic H_2S. The concentration of FeS at any given depth in the sediment, thus, is a result of formation by H_2S and oxidation back to limonite, as well as transformation to pyrite. Note that most pyrite is formed in the top 2 cm.

(1953) and Volkov (1961) to explain the formation of pyrite in the Black Sea. In addition, experimental studies (Rickard, 1969; Berner, 1960a) have shown that pyrite will form at low temperatures ($<100°C$) and neutral pH values characteristic of marine sediments by the reaction in H_2S solution of iron monosulfides with elemental sulfur.   By contrast, in the absence of elemental sulfur or of substances which produce it upon decomposition, there is little or no laboratory evidence for pyrite formation at low temperatures and neutral pH.   Furthermore, pyrite produced from elemental sulfur in the laboratory bears a striking textural resemblance to that found in natural sediments. It occurs as microconcretions, called framboids, which are microscopic spherical aggregates of submicron-sized pyrite crystals (see Fig. 10-6).

In some sediments black iron monosulfides are not converted to pyrite with depth.   An outstanding example is provided by older buried sediments of the Black Sea where black FeS-rich clays occur interbedded with gray pyritic clays containing no FeS minerals (Emel'yanov and Shimkus, 1962; Volkov, 1961).   The black clays are lower in total reduced sulfur and much lower in pyrite sulfur than the gray clays.   This is shown by the measurements of the author in Table 10-2.   Since the total reactive iron content of each sediment type is similar, the black clays must represent an arrested stage of diagenesis brought about by a paucity of H_2S and elemental sulfur (Volkov, 1961).   This was probably caused by a low concentration of SO_4^{--} and H_2S in the bottom waters at the time of deposition due to desalination of the Black Sea during the Wisconsin glacial maxima (Berner, 1970c).

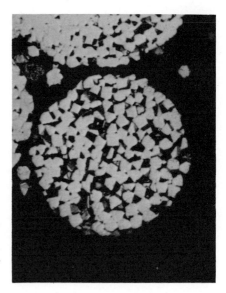

Fig. 10-6 Photomicrograph of an unusually large framboidal pyrite microsphere, showing the constituent crystals; Leicester member, Moscow fm., Middle Devonian of New York.   Reflected light ($\times$ 2000).

Table 10-2   Concentrations of monosulfide (acid soluble) sulfur, pyrite sulfur,
total reduced sulfur, total reactive iron, and organic carbon for interbedded black
and gray clays buried beneath the Black Sea
Total reduced sulfur includes that in FeS, Fe_3S_4, FeS_2, and S^0.   Total reactive iron represents
pyrite iron plus iron soluble in boiling 6 N HCl.   All concentrations are in % of $CaCO_3$-free,
HCl-soluble iron-free, dry weight

Depth in sediment, cm	Monosulfide S, %	Pyrite S, %	Total reduced sulfur, %	Total reactive Fe, %	% organic carbon
Black clays					
162–164	0.60	0.31	1.10	6.47	0.60
270–272	0.48	0.44	1.06	5.76	0.71
329–331	0.40	0.26	0.76	6.35	0.63
347–349	0.34	0.46	0.89	6.75	0.73
Gray clays					
166–168	0.01	1.99	2.02	6.42	0.59
239–241	0.01	1.82	1.85	6.31	0.52
297–299	0.01	1.68	1.71	6.11	0.58
362–364	0.02	1.56	1.61	5.60	0.87

REFERENCES

Baas Becking, L. G. M., Kaplan, I. R., and Moore, D., 1960, Limits of the natural environment in terms of pH and oxidation-reduction potentials, *Jour. Geol.*, v. 68, pp. 243–284.
Bell, D. L., and Goodell, H. G., 1967, A comparative study of glauconite and the associated
   clay fraction in modern marine sediments, *Sedimentology*, v. 9, pp. 169–202.
Berner, R. A., 1964a, Iron sulfides formed from aqueous solution at low temperatures and
   atmospheric pressure, *Jour. Geol.*, v. 72, pp. 293–306.
———, 1964b, Stability fields of iron minerals in anaerobic marine sediments, *Geochim. et
   Cosmochim. Acta*, v. 28, pp. 1497–1503.
———, 1969, Goethite stability and the origin of red beds, *Geochim. et Cosmochim. Acta*,
   v. 33, pp. 267–273.
———, 1970a, Sedimentary pyrite formation, *Am. Jour. Sci.*, v. 268, pp. 1–23.
———, 1970b, Low temperature geochemistry of iron, in *Handbook of geochemistry*,
   v. II-1, section 26, Springer-Verlag, Berlin.
———, 1970c, Pleistocene sea levels possibly indicated by buried black sediments in the
   Black Sea, *Nature*, v. 227, p. 700.
Burst, J. F., 1958, "Glauconite" pellets; their mineral nature and application to stratigraphic
   interpretation, *Am. Assoc. Petroleum Geologists Bull.*, v. 42, pp. 310–327.
Carroll, D., 1958, The role of clay minerals in the transportation of iron, *Geochim. et Cosmo-
   chim. Acta*, v. 14, pp. 1–27.
Cloud, P. E., 1955, Physical limits of glauconite formation. *Am. Assoc. Petroleum Geologists
   Bull.*, v. 39, pp. 484–492.

Doyle, R. W., 1967, *Eh and thermodynamic equilibrium in environments containing dissolved ferrous iron*, Ph.D. Dissertation, Yale University, New Haven, Conn., 100 p.

Emel'yanov, E. M., and Shimkus, K. M., 1962, Contribution to the study of variability of the deep sea sediments of the Black Sea, *Okeanologiya*, v. 2, pp. 1040–1049.

Fischer, A. G., 1963, Essay review of descriptive paleoclimatology, *Am. Jour. Sci.*, v. 261, pp. 281–293.

Garrels, R. M., 1959, Rates of geochemical reactions at low temperatures and pressures, in Abelson, P. H. (ed.), *Researches in geochemistry*, Wiley, New York, pp. 25–37.

———, and Christ, C. L., 1965, *Solutions, minerals, and equilibria*, Harper, New York, 450 p.

Hower, J., 1961, Some factors concerning the nature and origin of glauconite, *Am. Mineralogist*, v. 46, pp. 313–334.

Hutchinson, G. E., 1957, *A treatise on limnology*, v. I, *Geography, physics, and chemistry*, Wiley, New York, 1015 p.

Jedwab, J., 1967, Mineralization en greigite de debris vegetaux d'une vase recente (Grote Geul), *Soc. belge. geologie Bull.*, v. 76, pp. 1–19.

Kaplan, I. R., Emery, K. O., and Rittenberg, S. C., 1963, The distribution and isotopic abundances of sulphur in recent marine sediments off southern California, *Geochim. et Cosmochim. Acta*, v. 27, pp. 297–332.

Krynine, P. D., 1950, Petrology, stratigraphy, and origin of the Triassic sedimentary rocks of Connecticut, *Connecticut State Geol. and Natl. History Survey Bull. No. 73*, 247 p.

Langmuir, D., 1970, The effect of particle size on the reaction: hematite + water → goethite, *Geol. Soc. Amer. Abstracts of Annual Meetings*, pp. 601–602.

Ostroumov, E. A., 1953, Different forms of combined sulfur compounds in the bottom sediments of the Black Sea, *Akad. Nauk. SSSR, Inst. Okeanologii Trudy*, v. 7, pp. 70–90.

———, Volkov, I. I., and Fomina, L. C., 1961, Distribution pattern of sulfur compounds in bottom sediments of the Black Sea, *Akad. Nauk. SSSR, Inst. Okeanologii Trudy*, v. 50, pp. 93–124.

Polushkina, A. P., and Sidorenko, G. A., 1963, Melnikovite as a mineral species, *Zapiski Vses. Mineralogy. Obshch.*, v. 92, pp. 547–554.

Rickard, D. T., 1969, The chemistry of iron sulphide formation at low temperatures, *Stockholm Contr. Geology*, v. 20, pp. 67–95.

Schmalz, R. F., 1959, A note on the system Fe_2O_3–H_2O, *Jour. Geophys. Research*, v. 64, pp. 575–579.

Skopintsev, B. A., 1961, Recent work on the hydrochemistry of the Black Sea, *Okeanologiya*, v. I, pp. 243–250.

Smith, F. G., and Kidd, D. J., 1949, Hematite-geothite relations in neutral and alkaline solutions under pressure, *Amer. Mineralogist*, v. 34, pp. 403–412.

Van Houten, F. B., 1968, Iron oxides in red beds, *Bull. Geol. Soc. Amer.*, v. 79, pp. 399–416.

Van Straaten, L. M. J. U., 1954, Composition and structure of recent sediments of the Netherlands, *Leidse Geol. Mededel.*, v. 19, pp. 1–110.

Volkov, I. I., 1961, Iron sulfides, their interdependence and transformation in the Black Sea bottom sediments, *Akad. Nauk. SSSR, Inst. Okeanologii Trudy*, v. 50, pp. 68–92.

Walker, T. R., 1967, Formation of red beds in modern and ancient deserts, *Bull. Geol. Soc. Amer.*, v. 79, pp. 281–282.

Standard-state Thermodynamic Data for Common Sedimentological Substances

In the following table are presented values of the standard-state molar volume $v°$, entropy $s°$, enthalpy of formation from the elements $\Delta h_f°$, and Gibbs free energy of formation from the elements $\Delta F_f°$ for common chemical species of natural waters, sediments, and the atmosphere. All values are for 25°C, 1-atm total pressure, and the standard states as defined in Chapter 2. In the second column a number guide indicates the appropriate reference source listed at the end of the table. In addition, values for $v°$ are based on the data for ions of Owen and Brinkley (1941) and on the densities of minerals as found in Robie (1966) and in mineralogical texts. The reader should bear in mind that the quality of the data is highly variable and that accuracy is being constantly refined. For a more complete compendium which includes rarer substances, the reader is again referred to the references at the end of the table.

Composition	Reference	$v°$, $cm^3/mole$	$s°$, $eu/mole$	$\Delta h_f°$, $kcal/mole$	$\Delta F_f°$, $kcal/mole$
Gases					
O_2	1	24,400*	49.00	0	0
H_2	1	24,400	31.21	0	0
H_2O	1	24,400	45.11	−57.80	−54.64
CO_2	1	24,400	51.06	−94.05	−94.26
CO	1	24,400	47.30	−26.42	−32.81
CH_4	1	24,400	44.50	−17.89	−12.14
N_2	1	24,400	45.77	0	0
NH_3	1	24,400	46.01	−11.04	−3.98
SO_3	1	24,400	61.24	−94.45	−88.52
SO_2	1	24,400	59.40	−70.96	−71.79
H_2S	1	24,400	49.15	−4.82	−7.89

* For perfect gas.

Liquids					
H_2O	1	18.0	16.72	−68.32	−56.69

Dissolved solutes

			Neutral species		
$O_{2\,aq}$	2				4.08
$H_{2\,aq}$	2				4.22
$CO_{2\,aq}$	1		29.0	−98.69	−92.31
CO_{aq}	3				−28.70
$CH_{4\,aq}$	3				−8.29
$N_{2\,aq}$	3				4.38
$NH_{3\,aq}$	1		26.3	−19.32	−6.37
H_2S_{aq}	1		29.2	−9.4	−6.54
$H_2CO_{3\,aq}$	1		45.7	−167.0	−149.00
$H_4SiO_{4\,aq}$	4				−312.8
$H_2SO_{3\,aq}$	1		56	−145.5	−128.59
$H_3PO_{4\,aq}$	1		42.1	−308.2	−274.2
$NaHCO_3^0$	1				−202.56
$CaSO_4^0$	17				−312.84
$MgSO_4^0$	17				−289.23
$CaCO_3^0$	17				−262.93
$MgCO_3^0$	17				−239.62
$MgHPO_4^0$	16				−373.67
$CaHPO_4^0$	16				−397.53

Composition	Reference	$v°$, $cm^3/mole$	$s°$, $eu/mole$	$\Delta h_f°$, $kcal/mole$	$\Delta F_f°$, $kcal/mole$
Dissolved solutes (*continued*)					
		+1 *ions*			
H^+	1	0	0	0	0
Na^+	1	−1.5	14.4	−57.28	−62.59
K^+	1	8.7	25.5	−60.04	−67.47
NH_4^+	1	17.9	26.97	−31.74	−19.00
$Ca(OH)^+$	17				−171.72
$Mg(OH)^+$	17				−149.87
$CaHCO_3^+$	17				−274.38
$MgHCO_3^+$	17				−250.65
$Al(OH)_2^+$	5				−216.1
$Fe(OH)_2^+$	1				−106.2
$Fe(SO_4)^+$	16				−185.37
		+2 *ions*			
Ca^{++}	1, 9	−17.7	−13.2	−129.91	−132.35
Mg^{++}	9	−20.9	−32.7	−111.52	−108.76
Sr^{++}	1	−18.2	−9.4	−130.38	−133.2
Ba^{++}	1	−12.3	3	−128.67	−134.0
Fe^{++}	1		−27.1	−21.0	−20.30
Mn^{++}	1		−20	−53.3	−54.4
$Fe(OH)^{++}$	1		−23.2	−67.4	−55.91
$FeCl^{++}$	1		−22	−42.9	−35.9
		+3 *ions*			
Fe^{3+}	1		−70.1	−11.4	−2.52
Al^{3+}	1		−74.9	−125.4	−115.0
Mn^{3+}	1			−27	−19.6
		−1 *ions*			
F^-	1	−2.1	−2.3	−78.66	−66.08
Cl^-	1	18.1	13.17	−40.02	−31.35
OH^-	1	−5.3	−2.52	−54.96	−37.60
HCO_3^-	1	24	22.7	−165.18	−140.31
NO_3^-	1	29.3	35.0	−49.37	−26.43
NO_2^-	1		29.9	−25.4	−8.25
HSO_4^-	1		30.32	−211.70	−179.94
HSO_3^-	1		31.64	−150.09	−126.03
HS^-	1		14.6	−4.22	3.01

Composition	Reference	$v°$, cm³/mole	$s°$, eu/mole	$\Delta h_f°$, kcal/mole	$\Delta F_f°$, kcal/mole
Dissolved solutes (*continued*)					
			−1 ions		
$H_2PO_4^-$	1		21.3	−311.3	−271.3
$H_3SiO_4^-$	6				−299.3
$Al(OH)_4^-$	5				−311.3
$HFeO_2^-$	1				−90.6
$HMnO_2^-$	1				−120.9
MnO_4^-	1	43	45.4	129.7	−107.4
$NaSO_4^-$	18				−241.38
KSO_4^-	1				−246.11
$NaCO_3^-$	1				−190.54
			−2 ions		
CO_3^{--}	1	−3.7	−12.7	−161.63	−126.22
SO_4^{--}	1	14.5	4.1	−216.90	−177.34
SO_3^{--}	1		−7	−151.9	−116.1
$S_2O_3^{--}$	1		29	−154	−127.2
S^{--}	1				21.96
HPO_4^{--}	1		−8.6	−310.4	−261.5
			−3 ions		
PO_4^{3-}	1		−52	−306.9	−245.1
Solids					
			Silica and silicates		
SiO_2 quartz (α)	7	22.69	9.88	−217.7	−204.6
SiO_2 glass	4		11.2	−214.8	−203.1
Mg_2SiO_4 forsterite	7	43.67	22.8	−520.4	−491.9
Fe_2SiO_4 fayalite	7	46.39	34.7	−353.5	−329.4
$MgSiO_3$ clinoenstatite	7	31.47	16.2	−370.1	−349.4
$CaSiO_3$ wollastonite	7	39.94	19.6	−390.7	−370.3
$CaMgSi_2O_6$ diopside	7	66.10	34.2	−767.4	−725.8
$Ca_2Mg_5Si_8O_{22}(OH)_2$ tremolite	7	273	131.2	−2953.3	−2779.0
$KAlSi_3O_8$ microcline	10	109.5	52.5	−946.3	−892.8
$NaAlSi_3O_8$ low albite	10	100.2	50.2	−937.1	−884.0

Composition	Reference	$v°$, $cm^3/mole$	$s°$, $eu/mole$	$\Delta h_f°$, $kcal/mole$	$\Delta F_f°$, $kcal/mole$
Solids (*continued*)					
		Silica and silicates			
$KAl_3Si_3O_{10}(OH)_2$ muscovite	8	142	69.0	−1421.2	−1330.1
$Al_2Si_2O_5(OH)_4$ kaolinite	7	99	48.5	−979.5	−902.7
$Mg_3Si_4O_{10}(OH)_2$ talc	7	134.3	62.3	−1415.2	−1324.4
$NaAlSi_2O_6·H_2O$ analcite	7	97.5	56.0	−786.3	−734.1
		Carbonates			
$CaCO_3$ calcite	9	36.94	22.2	−288.59	−269.98
$CaCO_3$ aragonite	9	34.16	21.2	−288.65	−269.75
$MgCO_3$ magnesite	1	28.02	15.7	−266	−246
$FeCO_3$ siderite	1	29.38	22.2	−178.7	−161.1
$MnCO_3$ rhodocrosite	7	31.08	23.9	−212.4	−194.2
$SrCO_3$ strontianite	1	39.01	23.2	−291.2	−271.9
$BaCO_3$ witherite	1	45.81	26.8	−291.3	−272.2
$CaMg(CO_3)_2$ dolomite	9	64.35	37.1	−555.5	−516.6
$NaHCO_3·Na_2CO_3·2H_2O$ trona	1	105			−570.4
$NaHCO_3$ nahcolite	1	38.9	24.4	−226.5	−203.6
		Sulfates			
$CaSO_4$ anhydrite	11	45.94	22.0	−343.5	−315.69
$SrSO_4$ celestite	1	46.25	29.1	−345.3	−318.9
$BaSO_4$ barite	1, 16	52.11	31.6	−350.2	−325.0
$CaSO_4·2H_2O$ gypsum	17	74.31	46.36	−483.17	−429.36
Na_2SO_4 thenardite	7	53.34	35.73	−331.8	−303.7
$Na_2SO_4·10H_2O$ mirabolite	7	219.8	141.5	−1034.2	−871.5
		Phosphates			
$Ca_5(PO_4)_3OH$ apatite	7	159.7	93.3		
$Ca_5(PO_4)_3F$ apatite	7	157.6	92.7		
$FePO_4·2H_2O$ strengite	7	65	40.9	−451.5	−397.7
		Nitrates			
$NaNO_3$ soda niter	7	37.60	27.8	−111.54	−87.46
KNO_3 niter	7	48.04	31.81	−117.76	−93.91

Composition	Reference	$v°$, $cm^3/mole$	$s°$, $eu/mole$	$\Delta h_f°$, $kcal/mole$	$\Delta F_f°$, $kcal/mole$
Solids (*continued*)					
		Chlorides			
NaCl halite	7	27.02	17.33	−98.23	−91.81
KCl sylvite	7	37.53	19.70	−104.18	−97.52
		Oxides, hydroxides			
$Mg(OH)_2$ brucite	7	24.64	15.09	−221.2	−199.5
$Ca(OH)_2$ portlandite	7	33.06	19.93	−235.61	−214.67
TiO_2 rutile	7	18.80	12.04	−225.76	−212.5
$FeTiO_3$ ilmenite	7	31.71	25.3	−295.6	−277.1
Fe_2O_3 hematite	7	30.28	20.9	−196.8	−177.1
Fe_3O_4 magnetite	7	44.53	36.0	−267.4	−243.1
$HFeO_2$ goethite (limonitic)	12	23.4			−116.7
$Fe(OH)_3$ pptd.	1		23	−197.0	−166.0
$Fe(OH)_2$ pptd.	1		19	−135.8	−115.57
MnO_2 pyrolusite	7	16.61	12.7	−124.5	−111.3
MnO_2 birnessite	14				−108.3
MnO_2 nsutite	14				−109.1
Mn_2O_3 bixbyite	7	31.38	26.4	−229.2	−210.6
Mn_3O_4 hausmannite	7	46.96	35.5	−331.4	−306.0
$MnOOH$ manganite	14	19.2			−133.3
$Mn(OH)_3$ pptd.	1		23.8	−212	−181
$Mn(OH)_2$ pyrochroite	14	27.4			−147.34
Al_2O_3 corundum	7	25.57	12.18	−400.4	−378.1
$Al(OH)_3$ gibbsite	7	31.96	16.75	−306.4	−273.5
$AlOOH$ boehmite	7	19.54	11.58	−235.5	−217.7
$HAlO_2$ diaspore	15	17.76	8.43	−237.5	−218.6
		Sulfur and sulfides			
S orthorhombic	7	15.52	7.62	0	0
FeS pyrrhotite (troilite)	7	18.17	14.42	−24.2	−24.3
FeS mackinawite	13	20.5			−22.3
FeS pptd.	13				−21.3
FeS_2 pyrite	7	23.94	12.65	−41.0	−38.3
Fe_3S_4 greigite	13	72			−69.4
MnS alabandite	7	21.46	18.69	−49.0	−50.0

REFERENCES FOR TABLE

1. Garrels, R. M., and Christ, C. L., 1965, *Solutions, minerals, and equilibria*, New York, Harper, 450 p.
2. Free energies of formation calculated from the solubilities of H_2 and O_2 in water.
3. Thorstenson, D. C., 1969, *Equilibrium distribution of small organic molecules in natural waters*, Unpublished Ph.D. Thesis, Northwestern University, Evanston, Ill.
4. Calculated from values given in reference 1; corrected for the more recent ΔF_f° value for quartz given in reference 7.
5. Reesman, A. L., Pickett, E. E., and Keller, W. D., 1969, Aluminum ions in aqueous solution, *Am. Jour. Sci.*, 267, pp. 99–113.
6. Calculated from the data of reference 4, and the first dissociation constant of H_4SiO_4.
7. Robie, R. A., 1966, Thermodynamic properties of minerals, in Clark, S. P., Jr. (ed.), *Handbook of physical constants*, Geol. Soc. Mem. 97, pp. 437–458.
8. Barany, R., 1964, *U.S. Bureau of Mines Report of Investigations* 6356.
9. Langmuir, D., 1964, Stability of carbonates in the system $CaO–MgO–CO_2–H_2O$, Unpublished Ph.D. Thesis, Harvard University, 142 p.; Langmuir, D., 1968, Stability of calcite based on aqueous solubility measurements, *Geochim. et Cosmochim. Acta*, v. 32, pp. 835–851.
10. Waldbaum, D. R., 1968, High-temperature thermodynamic properties of alkali feldspars, *Contr. Mineral. and Petrol.*, v. 17, pp. 71–77.
11. Values calculated from the data of Hardie (Hardie, L. A., 1967, The gypsum-anhydrite equilibrium at 1 atm pressure, *Am. Mineralogist*, v. 52, pp. 171–200) and data for gypsum from reference 17.
12. Calculated from the data of Berner (Berner, R. A., 1969, Goethite stability and the origin of red beds, *Geochim. et Cosmochim. Acta*, v. 33, pp. 267–273.) and ΔF_f° of hematite and water. The ΔF_f° value pertains only to limonitic goethite with $r \approx 1000$ Å.
13. Berner, R. A., 1967, Thermodynamic stability of sedimentary iron sulfides, *Am. Jour. Sci.*, v. 265, pp. 773–785.
14. Bricker, O. P., 1965, Some stability relations in the system $Mn–O_2–H_2O$ at 25°C and one atmosphere total pressure, *Am. Mineralogist*, v. 50, pp. 1296–1354.
15. Value of s° from reference 7; value of ΔF_f° from Reesman, A. L., and Keller, W. D., 1968, Aqueous solubility of high-alumina and clay minerals, *Am. Mineralogist*, v. 53, pp. 929–942; value for ΔH_f° calculated from $\Delta F_f^\circ + T\Delta s_f^\circ$.
16. Sillén, L. G., 1964, *Stability constants of metal-ion complexes*, Section I. *Inorganic ligands*, London, the Chemical Soc., pp. 1–356.
17. Calculated from values given in reference 1 corrected for ΔF_f° values for Ca_{aq}^{++} and Mg_{aq}^{++} given in reference 9.
18. Calculated from the data of Pytkowicz, R. M., and Kester, D. R., 1969, Harned's rule behavior of $NaCl–Na_2SO_4$ solutions explained by an ion-association model, *Am. Jour. Sci.*, v. 267, pp. 217–229.

Values of v° for ions are taken from Owen, B. B., and Brinkley, S. R., 1941, Calculation of the effect of pressure upon ionic equilibria in pure water and in salt solution, *Chem. Reviews*, v. 29, pp. 461–474.

Appendix II

Debye-Hückel Equation Parameters

The Debye-Hückel expression for individual ion activities, as discussed in Chapter 2, is

$$\log \gamma = \frac{-AZ^2\sqrt{I}}{1 + åB\sqrt{I}}$$

In this equation parameters A and B at 1-atm pressure as a function of temperature (from Manov et al., 1943) are

Temperature °C	A	B($\times 10^{-8}$)
0	0.4883	0.3241
5	0.4921	0.3249
10	0.4960	0.3258
15	0.5000	0.3262
20	0.5042	0.3273
25	0.5085	0.3281
30	0.5130	0.3290
35	0.5175	0.3297
40	0.5221	0.3305
50	0.5319	0.3321
60	0.5425	0.3338

Values of the ion-size parameter $\mathring{a}$ for common ions encountered in natural water (adapted from Klotz, 1950, p. 331) are

$\mathring{a} \times 10^8$	Ion
2.5	NH_4^+
3.0	K^+, Cl^-, NO_3^-
3.5	OH^-, HS^-, MnO_4^-, F^-
4.0	SO_4^{--}, PO_4^{3-}, HPO_4^{--}
4.0–4.5	Na^+, HCO_3^-, $H_2PO_4^-$, HSO_3^-
4.5	CO_3^{--}, SO_3^{--}
5	Sr^{++}, Ba^{++}, S^{--}
6	Ca^{++}, Fe^{++}, Mn^{++}
8	Mg^{++}
9	H^+, Al^{3+}, Fe^{3+}

REFERENCES

Klotz, I. M., 1950, *Chemical thermodynamics*, Prentice-Hall, Englewood Cliffs, N.J., 369 p.
Manov, G. G., Bates, R. G., Hamer, W. J., and Acree, S. F., 1943, Values of the constants in the Debye-Hückel equation for activity coefficients, *Jour. Am. Chem. Soc.*, v. 65, pp. 1765–1767.

Derivation of Useful Thermodynamic Equations

Equations useful to chemical sedimentology can be derived from a few basic chemical thermodynamic equations. The basic equations are here classified into four categories: (1) the combined first and second laws of thermodynamics; (2) definitions of enthalpy and Gibbs free energy; (3) definition of the activity; (4) expressions for work. For the purpose of this Appendix the equations presented under categories 1 to 4 can be considered simply as definitions. From them the equations employed in studying sediments are obtained by straightforward mathematical derivation combined with the criterion that the change in free energy for a chemical reaction at equilibrium is equal to zero. The basic equations follow.

1. *Combined first and second laws of thermodynamics*

$$dU = T\,dS - dW + \mu_i\,dn_i + \mu_j\,dn_j + \cdot\,\cdot\,\cdot \tag{1}$$

where U = internal energy

$\quad\quad S$ = entropy

$\quad\quad T$ = absolute temperature

W = work

μ = chemical potential

n = number of moles of component i, j, etc.

2. *Definitions of enthalpy and Gibbs free energy*

$$H \equiv U + PV \tag{2}$$

$$F \equiv H - TS \tag{3}$$

where H = enthalpy

P = pressure

V = volume

F = Gibbs free energy (or Gibbs function)

3. *Definition of the activity*

$$\mu_i = \mu_i^\circ + RT \ln a_i \tag{4}$$

where μ_i° = chemical potential in a standard state

a = activity

R = gas constant

4. *Expressions for work*

$$dW = P\,dV \qquad \text{(mechanical work)} \tag{5}$$

$$dW = E\,dZ \qquad \text{(electrical work)} \tag{6}$$

where E = electrical potential

Z = charge

Now from Eqs. (2) and (3)

$$dF = dU + P\,dV + V\,dP - T\,dS - S\,dT \tag{7}$$

Substituting Eqs. (1) and (5) in Eq. (7)

$$dF = V\,dP - S\,dT + \mu_i\,dn_i + \mu_j\,dn_j \;\cdots \tag{8}$$

Therefore

$$\left.\frac{\partial F}{\partial P}\right|_{T,\,n_i,\,n_j\,\cdots} = V \tag{9}$$

$$\left.\frac{\partial F}{\partial T}\right|_{P,\,n_i,\,n_j\,\cdots} = -S \tag{10}$$

$$\left.\frac{\partial F}{\partial n_i}\right|_{P,\,T,\,n_j\,\cdots} = \mu_i \tag{11}$$

By double differentiation it can be also seen that

$$\frac{\partial \mu_i}{\partial P}\bigg|_{T,\, n_i,\, n_j} = v_i \tag{12}$$

$$\frac{\partial \mu}{\partial T}\bigg|_{P,\, n_i,\, n_j} = -s_i \tag{13}$$

where $v_i = \dfrac{dV}{dn_i}\bigg|_{P,\, T,\, n_j \,\cdots}$ \hfill (14)

$$s_i = \frac{dS}{dn_i}\bigg|_{P,\, T,\, n_j \,\cdots} \tag{15}$$

also

$$h_i = \frac{dH}{dn_i}\bigg|_{P,\, T,\, n_j \,\cdots} \tag{16}$$

Equation (8) at constant P and T can be integrated at constant μ (since F is a state function) to give

$$F = \mu_i n_i + \mu_j n_j + \cdots \tag{17}$$

For a chemical reaction the total free-energy change ΔF is equal to the free energy of the products minus the free energy of the reactants. For a generalized reaction of the type

$$aA + bB \rightarrow cC + dD$$

at constant P and T, the free-energy change *per mole* from Eq. (17) is

$$\Delta F = c\mu_C + d\mu_D - (a\mu_A + b\mu_B) \tag{18}$$

From Eqs. (4) and (18)

$$\Delta F = c\mu_C^\circ + d\mu_D^\circ - (a\mu_A^\circ + b\mu_B^\circ) + RT\ln\left(\frac{a_C{}^c\, a_D{}^d}{a_A{}^a\, a_B{}^b}\right) \tag{19}$$

Setting

$$\Delta F^\circ = c\mu_C^\circ + d\mu_D^\circ - (a\mu_A^\circ + b\mu_B^\circ) = \Delta\mu^\circ \tag{20}$$

and

$$Q = \frac{a_C{}^c\, a_D{}^d}{a_A{}^a\, a_B{}^b} \tag{21}$$

$$\Delta F = \Delta F^\circ + RT\ln Q \tag{22}$$

For a chemical reaction to proceed spontaneously as written the value of ΔF calculated from Eq. (22) must be less than zero. Otherwise it can proceed only in the opposite direction. This is a consequence of the second law of thermodynamics. At equilibrium no net reaction occurs and thus, $\Delta F = 0$ and $Q = K$, the equilibrium constant. Therefore, for equilibrium

$$\Delta F^\circ = -RT \ln K \tag{23}$$

This is the important relation linking the equilibrium constant used in geochemical calculations to a fundamental thermodynamic variable, the free energy.

The pressure effect on K can be derived directly from Eqs. (12) and (23) to obtain

$$\left. \frac{d \ln K}{dP} \right|_{T, n_i, n_j, \dots} = \frac{-\Delta v^\circ}{RT} \tag{24}$$

The temperature effect on K can be derived from a modification of Eq. (3). At constant P and T from Eq. (3)

$$\Delta F = \Delta H - T \Delta S \tag{25}$$

or

$$\Delta F^\circ = \Delta h^\circ - T \Delta s^\circ \tag{26}$$

From Eq. (23)

$$\left. \frac{\partial \ln K}{\partial T} \right|_{P, n_i, n_j \dots} = \frac{\Delta F^\circ}{RT^2} - \frac{1}{RT} \left. \frac{d \Delta F^\circ}{dT} \right|_{P, n_i, n_j \dots} \tag{27}$$

From Eq. (13)

$$\left. \frac{\partial \Delta F^\circ}{\partial T} \right|_{P, n_i, n_j \dots} = -\Delta s^\circ \tag{28}$$

Therefore, from Eqs. (26), (27), and (28)

$$\left. \frac{\partial \ln K}{\partial T} \right|_{P, n_i, n_j \dots} = \frac{\Delta h^\circ}{RT^2} \tag{29}$$

Chemical reactions in which oxidation and reduction take place can be utilized to accomplish electrical as well as mechanical work. For a thermodynamically reversible redox reaction the change in free energy at constant P and T can be equated with the electrical work done. Thus, from Eq. (6)

$$dF = E \, dZ \tag{30}$$

Setting

$$dZ = \mathscr{F} \, dn \tag{31}$$

where n = number of molar equivalents of electrons transferred during the redox reaction

$\mathscr{F}$ = the Faraday constant

Upon substitution and integration

$$\Delta F = n \mathscr{F} E \tag{32}$$

also

$$\Delta F^\circ = n \mathscr{F} E^\circ \tag{33}$$

Substituting Eqs. (32) and (33) in Eq. (22) and solving for E:

$$E = E^\circ + \frac{RT}{n\mathscr{F}} \ln Q \tag{34}$$

This is an important relation linking the voltage developed by a redox reaction in an electrochemical cell to the ratio of the activities of products to reactants. As Q approaches K, ΔF and thus E approach zero. Although this book is not concerned with electrochemical cells, E can still be employed conceptually as a measure of the deviation from redox equilibrium. Also, in the form of Eh it can be used as a measure of the redox state of a solution relative to a standard redox state. It is in the latter context that E is most commonly used in chemical sedimentology.

In summary, six fundamental equations applicable to chemical sediment-ology as derived above are:

$$\Delta F = \Delta F^\circ + RT \ln Q \tag{22}$$

$$\Delta F^\circ = -RT \ln K \tag{23}$$

$$\left. \frac{\partial \ln K}{\partial P} \right|_{T, \, n_i, \, n_j \, \cdots} = \frac{-\Delta v^\circ}{RT} \tag{24}$$

$$\left. \frac{\partial \ln K}{\partial T} \right|_{P, \, n_i, \, n_j \, \cdots} = \frac{\Delta h^\circ}{RT^2} \tag{29}$$

$$\Delta F^\circ = n \mathscr{F} E^\circ \tag{32}$$

$$E = E^\circ + \frac{RT}{n\mathscr{F}} \ln Q \tag{34}$$

Name Index

Subject Index